THE TRI-ANG, MINIC & SPOT-ON PRICE GUIDE

Frank Thompson

Ernest Benn

First published 1984 by Ernest Benn Limited
Sovereign Way, Tonbridge, Kent TN9 1RW
©Frank Thompson 1984
Cover design and illustration by Martin Lee
ISBN 0 510 00171 8
Printed in Great Britain

CONTENTS

INTRODUCTION

In this price guide you will find a comprehensive range of British and overseas railway models plus the exciting range of Minic clockwork and Spot-On diecast toys which have fascinated collectors in almost every country you care to name.

This book, I believe, is the first publication to give prices to the trains and accessories made by one of the world's greatest toy companies, Lines Brothers, who also brought great excitement into the world of collecting in 1959, when they opened the premises of an old toy firm in Ireland. The name was Spot-On and it meant what it said, with a perfect design and scale which had never been accomplished before in the era of toy making. With their famous triangular trade mark, Lines Brothers made it their policy to satisfy model enthusiasts and toy collectors with an ever increasing range of top quality railway sets and other toys which were second to none in design and execution. It was said once that the Triang-Minic works became as famous as Clapham Junction as the company was handling almost two thousand trains a day.

As with previous titles in this series, the models list is accompanied by full details of colours, dates of issue and deletion, original prices, length etc. Against each model are the prices that a collector should expect to pay for them. The prices have been broadly graded as mint boxed, mint unboxed and good condition. It should be noted that some models were never boxed in which cases prices are given only for good and fair condition. There is also a section concerning boxes and catalogues, both areas of interest for the serious collector. I have tried to make this book as comprehensive and accurate as possible but the possibility of error always exists and so I would welcome hearing from other collectors any comments or suggestions they may have.

I would like to thank all the people who have helped me with this publication, above all my wife Anne who was responsible for typing not only this book but the others in the series.

THE LINES BROTHERS AND TRI-ANG

One of the great success stories is that of Walter Lines and his two brothers. After the First World War they decided not to return to the family business where they had worked since their school days preparing cow tails and horse hair for rocking horses.

They put their money together, plus some savings which their mother had kept for them, and bought large premises near a wharf in the Old Kent Road with some showroom space in the City of London itself. They organized one of the first mass production lines to make rocking horses, dolls' houses, prams, fairy cycles, scooters, pedals cars, large pull-along railway engines and what was to become perhaps their most famous product, clockwork toys.

Whereas Continental firms were charging pounds for their clockwork toys, Lines Brothers made high quality moving models for only a few pennies. This was the real reason for their success, and the word rapidly spread. The toys were not only made to be played with but they were collectable and affordable. At last working class children had been catered for.

The famous triangle trade mark began to appear on toys in 1924 when it was registered and incorporated the name 'Triangtois', intended purposely to rhyme with 'The World's Best Toys'. By 1926 the Lines brothers had built the largest toy factory in the world at Merton, near Wimbledon and in 1927 the more familiar triangle trade mark was registered and by 1931 the name Triang, correctly spelt Tri-ang, had been registered as a trade mark. Shortly after this the name Minic was created by Walter Lines from Mini, meaning small, and 'c' standing for clockwork and was registered as a trade mark in 1935. Soon toys were being exported to over forty countries and the company was selling more in the year 1936 than all the other similar companies in

Great Britain put together. This record was even enhanced in 1937, 1938 and 1939.

It is impossible to say how many models the Triang Minic company had made by December 1939 when officially all trade was stopped and models deleted. A few unofficial items were produced up to 1941, though none of them ever got into the shops. They were brought out of the factory by workmen and sold privately. This is why the odd one-off and rare item will appear in a collection.

All though the war years Lines brothers continued to advertise while no actual models were shown. It was simply to keep the name of Triang before the public. Curiously enough although before the war Lines brothers had refused to make toy weapons, during the war the company was responsible for the manufacture of the Sten Gun.

Factories were built and acquired in Birmingham, South Wales and certain parts of the North-East. They were producing scale model aircraft, clockwork toys, dolls, teddy bears and all the accessories that went with them. Even a paint making factory was bought to ensure that the highest standards were maintained in authentic colour. Hamley's of Regent Street London also became part of the ever growing Lines Empire.

After the war production started again in 1945 and it seemed that the public was as keen as ever to buy Triang products. As the company announced in 1945/46. 'For some time past you have only seen the trade marks but we hope it will not be long before the world famous toys appear once again in your local shop or toy-store.'

By Christmas 1946 many metal and tinplate Minic toys were back in the shops, although it was well into 1947 before supplies returned to normal. By 1948 plastic models began to appear, which many people regarded as a mistake, especially the keen collector like myself. However, in 1964 the Triang group took over Meccano, including Hornby Dublo and Dinky toys, which gave them a virtual monopoly of model train production and the catalogues began to carry the joint names of Triang and Hornby, with, from 1968 the addition of Minic as well. I must point out that after securing Dinky models in 1964, the Spot-On items began to disappear from the scene. There also was the ever constant fear of fires, bombings and other upsets through the troubles in Northern Ireland, therefore Spot-On was eventually phased out.

In 1971 the word Minic was dropped from the front of the catalogue and in January 1972 the ailing Rovex company, who owned Triang, were bought by the Dunbee-Combex-Marx organisation. At the end of 1979 the Triang part of the company was bought out by Morris Vulcan who, in turn, went into the hands of the liquidator but in 1983 the name Triang was sold again and now continues as Sharna-Triang. For collectors the golden age of the company ended around 1973 but it is good to see that the name lives on.

SPOT-ON FOR PERFECTION

Although the lifespan of Spot-On models was comparatively short, their diecast models nonetheless rank amongst the best in the world.

With a factory in Castlereagh Road, Belfast, Spot-On began full production in 1959 as a very important part of the Triang group. The first catalogue was only in black and white and very limited, issued in late 1958 as advance publicity. This is something of a rarity and I advise any collector to hold on to such a possession. By 1961 Triang were producing a beautiful full colour catalogue. The models were based on every-day vehicles, often in the actual colours of the full sized version, although it is important to note that colours can vary from those shown in the actual catalogue. Beautifully made and detailed they rapidly became collectors items and interest in them has continued long after the company stopped manufacturing in the mid-sixties. In fact good Spot-On models are frequently more valuable than Dinky or Matchbox toys of the same period. The reason for Spot-On's demise is that the Irish factory had suffered continuously from fires, bombings and strikes and although plans were made to move the manufacturing process to Binns Road, Liverpool, this never materialised.

But during the brief life of the company many remarkable models were produced from the first, numbered 100, which was the Ford Zodiac through to the extremely rare Mulliner coach. Other rare models include a Rolls Royce set in the Royal authentic Maroon or dark Red livery complete with figures of the Queen and Prince Philip, a savings Bank Van and a Sailing Dinghy. Commercial vehicles, in particular, are of special interest and value to collectors.

I have endeavoured to make this list complete and I would welcome any comments about other models I perhaps have missed which other collectors may own. Please send a S.A.E. with any letters, which I assure you will be answered regardless of how busy I may happen to be.

HINTS FOR COLLECTORS

The best place to find a collector or to start your own collection is a Swapmeet or Fleamarket. Sometimes they are a combination of the two. Although the major auction houses such as Sotheby and Christie's are now only too willing to sell diecast toys, most business is still conducted at Toy Collectors' Fairs which have sprung up all over Europe and the United States. Many have become established all over Europe and the United States. Many have become established as annual events, some are monthly and a few are even weekly. In 1976/7 there might have been about 400 Swapmeets a year, now there are probably 400 a month in the United Kingdom alone.

National and particularly local newspapers give details of Swapmeets and Fleamarkets all the time. The Collectors' Gazette, Exchange and Mart, local radio and telelvision as well, of course, as collector's shops are all useful sources of information.

Another, and perhaps unexpected, place which attracts collectors is the Traction Engine Rally. At every Traction Engine Rally there are trade stands with dealers buying and exchanging models. Many are special promotional models, advertising a particular show. A publication called 'The World's Fair' lists all the Traction Engine Rallies to be held.

Similarly, the major Agricultural Shows attract many model enthusiasts as do meetings held by various car clubs and Bus and Coach preservation societies. Information on meetings is usually available locally. The possibility of buying a special promotional model at some of these events is an added incentive for attending.

It is as well to remember that a collector, should, wherever possible, buy an additional model as these can become exchangeable and thus a better means of finding those items which may be necessary to complete a collection.

GENERAL HINTS

If a toy has a box with it, always keep the item boxed at all times and look after the box as well as you would look after the toy. As you will see from the empty box section they too are worth money. If you intend to put your collection of models in a glass case or cabinet, make sure that the area is free from damp or excesses of heat or cold. One of the worst things for devaluing a model can be sunshine as it can fade the colour or even parts of the model and cause it to wear. Always place the model on top of the box in the showcase. When you transport models always take time to wrap up and pack them neatly, remembering to place sufficient packing between each layer of models. Never use newspaper to wrap models, not even on the outside as newsprint always seems to find its way onto the models or boxes to cause serious devaluation. Use a soft tissue paper or even soft toilet rolls.

Many models, such as Dinky and Matchbox, cannot readily be taken apart and put back together again whereas Minic models can. It is possible to take them apart and re-assemble them in what seems an authentic way, yet the wrong body colour can easily get mixed up. This can lead unsuspecting collectors to believe that they have found rare models so always check with an expert, especially if you are being asked to pay a high price for a particular model.

If you find an old toy in the attic or wherever, under no circumstances should you try and clean it. It is very easy to break or damage them and many people have completely destroyed old dolls by mistakenly thinking they would be better for a wash. Never wash clothes or outfits of any kind attached to a toy. Leave them and consult an expert. Remember the old saying 'Where there's muck there's money'.

A NOTE ON THE PRICES

Any price guide is certain to cause controversy and in particular one to an area of collecting where prices are often rising very rapidly. There will always be variations of opinion and you may find that a model which is selling for £50 in one place is available for £5 in another. This is not as true as it was but I hope very much that this guide will set a general standard to be followed.

A fine example with regard to the varying prices is the Belgian market where prices are very much higher than in other parts of Europe. If the English collector can afford a holiday in Belgium it would pay him to visit that part of the world where I found nothing but hospitality and civility in the best tradition.

The prices were as accurate as possible at the time of going to press but variations may have occurred in the months it has taken for the book to be produced. Often an item is as valuable as the person can afford to pay and this is particularly true of rare models where cost can become unimportant to the buyer. Nonetheless I would welcome comments from any collector, not only on prices, but on any other aspect of the guide.

EXPLANATIONS

To make it as easy as possible for the collector to find his or her model the book has been set out in sections, each devoted to a particular class or type of product. Within each section the models have been listed in numerical order, where appropriate, with all the necessary details of colour, tyres, etc. Each entry gives the date a model was issued, deleted or had its number changed for some reason. Where no deletion date is given it should be assumed that the model is currently available.

Prices are listed for Mint Boxed, Mint Unboxed and Good Condition. In the case of boxed sets only a Mint Boxed price is given.

Lengths are given in millimetres.

The price given in the descriptive section is the original price at which the model was sold.

In the case of the models intended for overseas markets the original selling price given is a sterling equivalent.

IMPORTANT NOTICE

The prices given in this guide are the prices that you should expect to pay in order to buy an item. They are not necessarily what you should expect to receive when selling to a dealer. Although every care has been taken in compiling this price guide, neither the publisher nor the author can accept any responsibility whatsoever for any financial loss or other inconvenience that may result from its use.

The correct spelling of Triang is Tri-ang but since the one word version is in common use it was felt reasonable to follow this usage and drop the hyphen.

MINIX SERIES

To compete with the Dinky Dublo cars and the small Matchbox 1-75 series the Triang Company brought out a series of Minix models which were exactly like the Dinky Dublo series in size. There were approximately 12 to 15 models consisting of a Post Office van in red, a taxi in blue, a taxi in black, a lorry in green and a Minix delivery van in blue with red lettering. Also a tractor in dark grey and a milk delivery van in cream with blue lettering. There was a sports car in red, a cattle truck in green and a log carrying vehicle in grey. There was a telephone van in green, a flat wagon in blue and a milk float with a small brown horse and the milk float was in cream and blue. There were possibly other models and colours could vary. Made approximately between 1961 and 1966 and were sold for 1/6d. in old coinage. They were mainly produced with the railway models in mind and they are quite rare. I have seldom seen more than six in any one collection. These models are mentioned from time to time in certain catalogues and Triang Magazines.

ABBREVIATIONS
MB Mint Boxed
MU Mint Unboxed
GC Good Condition
FC Fair Condition

LOCOMOTIVES

MODEL	MB	MU	GC

R50. 4-6-2 Princess Elizabeth.
Black. 7P Pacific electric express loco and the first model ever
produced by the Triang Company in 1955. Deleted in 1970. Price
39/6d. Matching tender for this loco was R30 in same livery. Loco
17.8cms. Tender 8.9cm. — £40 — £35 — £25

R52. 0-6-0 Electric Tank Loco.
Black livery. Price 37/6d. Issued 1956. Deleted 1970. 12.1cms. — £15 — £12 — £10

R53. 4-6-2 Princess Elizabeth.
Green livery. Fitted with Walschaerts gear. 17.8cms. Number of
tender was R31 in matching livery. Price 56/-. Issued 1966. Deleted
1970. 8.9cms. — £40 — £35 — £30

R55. Diesel Loco.
Silver grey, blue and red with Triang Railway decals. Price 42/-.
Issued 1956. Deleted 1970. 20cms. — £15 — £10 — £7

R55 CN. Canadian National Diesel.
Streamlined Canadian Diesel Engine in red, black and grey with
number '4008' decal. Complete with working headlight and knurled
steel driving wheel. Price 75/-. Issued 1967. Deleted 1973. 24cms. — £9 — £7 — £5

R56. Electric Tank Loco
Black. Price 39/6d. Issued 1955. Deleted 1970. 17.5cms. — £20 — £15 — £10

R57. Diesel Dummy End Loco.
Silver grey, blue and red with Triang Railway decals. Made
specifically to go with R55 Diesel Loco. Price 25/-. Issued 1956.
Deleted 1970. 20cms. — £9 — £7 — £5

R58. Diesel B Unit.
Specifically made to go with R55 and R57. Silver grey, blue and
red with matching Triang Railway decals in yellow. Price 21/-.
Issued 1956. Deleted 1970. 20.3cms. — £10 — £7 — £5

R59. Tank Loco.
Black with red lines and Lion decal with the number '82004' on
sides. Price 52/6d. 17.6cms. — £20 — £15 — £10

R138. Snow Plough.
Green and black with white decals and emblems. Opening side
wings. Price 40/5d. Issued 1962. Deleted 1970. 12.2cms. — £15 — £10 — £7

R150. 4-6-0 Class B12 Locomotive.
Black livery with B.R. decals and number '61573' on the sides.
Matching tender is R39. Price with smoke 47/-. Without smoke
39/6d. Issued 1962. Deleted 1970. Loco 15.8cms. Tender 8.2cms. — £20 — £15 — £10

R151. 0-6-0 Saddle Tank Loco.
Black with B.R. decals and number 748. A clockwork model. Price
31/-. Issued 1956. Deleted 1970. 12.2cms. — £15 — £12 — £9

R152. 0-6-0 Diesel Shunter.
Black livery with 13002 on sides. Price 35/-. Issued 1956. Deleted
1970. 12.2cms. — £15 — £12 — £9

MODEL	MB	MU	GC
R153. 0-6-0 Saddle Tank Loco. Black with B.R. decals and '748' on sides. Same model as 151 only electric. Price 37/-. Issued 1955. Deleted 1970. 12.2cms.	£15	£12	£9
R159. 0-6-0 Diesel Shunter. Black with number '13002' on sides. Clockwork model. Price 27/6d. Issued 1956. Deleted 1970. 12.2cms.	£12	£10	£8
R155. Diesel Switcher. Yellow body with black chassis and black roof on cab. Triang Railways decals in white on red background and number '4864' on cab sides. Price 35/-. Issued 1957. Deleted 1970. 20cms.	£20	£15	£10
R156. S.R. Suburban Motor Coach. Lime green with silver grey roof and black chassis and wheels. Price 39/6d. Issued 1959. Deleted 1970. 22.7cms.	£25	£20	£15
R157. Diesel Power Car. Rich medium or dark green power car with grey or silver roof and silver or black interior, complete with red or yellow lining. Made with seats and four destination boards. Price 46/-. Issued 1964. Deleted 1970. 22.7cms.	£20	£15	£10
R158. Diesel Trailer Car A non-powered car in matching livery to go with R157. With seats and four destination boards. Price 13/-. Issued 1957. Deleted 1970. 22.7cms.	£6	£5	£3
R159. Double Ended Diesel Loco. Silver grey, orange, yellow and red with black chassis and roof cover. A streamlined loco with working headlights. Only a limited number made in this livery. Price 56/-. Issued 1964. Deleted 1966. 22cms.	£75	£60	£55
R159/B. Double Ended Diesel Loco. Blue, lemon and black with silver horns and working headlights. Price 55/-. Issued 1964. Deleted 1970. 22cms.	£15	£10	£7
R225. S.R. Suburban Motor Coach. A non-powered coach with built-in seat unit to go with 156 and came in same green, black and grey matching livery. Price 15/-. Issued 1962. Deleted 1970. 22.7cms.	£10	£7	£5
R251. 0-6-0 Class 3F Loco. Black. Tender is R33 in matching livery. Price 47/6d. Issued 1962. Deleted 1970. Loco 11.4cms. Tender 8.7cms. An RT 520 Smoke Unit could be fitted to this locomotive.	£20	£15	£10
R253. 0-4-0 Dock Shunter Red and black with working headlight. Price 32/6d. Issued 1962. Deleted 1970. 10cms.	£10	£7	£5
R254. 0-4-0 Steeple Cab Loco. Green with T.R. decals in yellow and red. Operating pantograph. Price 29/6d. 10.5cms.	£15	£12	£9
R255. 0-6-0 Saddle Tank Loco. Black and green or black and blue, worth double. A clockwork loco. Price 25/-. Issued 1962. Deleted 1970. 12.2cms.	£15	£10	£7

MODEL	MB	MU	GC

R257. Double Ended Electric Loco.

Green, grey, black with gold linage and Transcontinental decals.
Operating pantographs and working headlights. Works from
overhead power supply system. Price 39/6d. Issued 1962. Deleted
1970. 22cms.

£20 £15 £10

R258/S. 4-6-2 The Princess Royal Loco.

Maroon with smoke and Magnadhesion, complete with Walschaerts
valve gear. Price 75/-. Issued 1961. Deleted 1970. Matching tender
is R34. 17.8cms.

£35 £30 £25

R259S. 4-6-2 'Britannia' Class 7P6F. Loco.

Green livery with Walschaerts Valve Gear with smoke and
Magnadhesion. Price without smoke 56/-. With smoke 75/-. Issued
1962. Deleted 1970. Tender is R35. Loco 28.4cms. Tender 9.8cms.

£35 £30 £25

R346. Stephenson's Rocket Train.

Yellow and black with authentic decals in black and red. With
crew, smoke and Magnadhesion. Price 49/6d. Issued 1964. Deleted
1970. Loco 5.6cms. Tender 3.8cms.

£50 £45 £35

R350. 4-4-0 L1 Class 3P Loco.

Green with smoke and Magnadhesion. Price 42/-. Issued 1962.
Deleted 1970. Loco 13.5cms. Tender was R36, 8.7cms.

£25 £20 £15

R351. Co-Co Class AM2 Electric Loco.

Green with twin operating pantographs with Magnadhesion
operating from the overhead power supply system. Price 63/-.
Issued 1962. Deleted 1970. 23.7cms.

£30 £25 £20

R352. Budd Rail Diesel Car.

Medium or dark silver grey, red and yellow. Magnadhesion and
built-in interior fittings. Price 52/6d. Issued 1962. Deleted 1970.
25.6cms.

£25 £20 £15

R353. Yard Switcher.

Yellow and black with working headlights. Price 33/6d. 8.7cms.

£12 £9 £7

MODEL	MB	MU	GC

R354. 'Lord of the Isles' Locomotive.

Great Western Railway livery of green, black, brown and yellow, this Dean Single Locomotive came with smoke and Magnadhesion. R354S has smoke. Price without smoke, 52/6d. With smoke 77/-. Issued 1962. Deleted 1970. Matching tender is R37. Loco 13.3cms. Tender 10.3cms.

	£35	£30	£25

R355. 0-4-0 Industrial Loco, 'Connie'.

Blue with black cab roof, front, chassis etc. Model also available as 'Nellie' No. 7. Price 32/6d. Issued 1962. Deleted 1973. One of the most reliable locos ever made. 10.5cms.

	£15	£10	£7

355.Y, Industrial loco 'Connie' No.6 in yellow. Otherwise as above.

	£15	£10	£7

355.Y, Industrial loco 'Polly' No.9 in red. Issued 1964. Otherwise as above.

	£12	£8	£6

R356. 4-6-2 Battle of Britain Class 7P5F Loco.

Green combined with black and grey and showing 'Winston Churchill' decals and No. '34051'. Smoke and Magnadhesion. Price 67/6d. Issued 1962. Deleted 1970. Matching tender is R38. Loco 17.4cms. Tender 9.5cms.

	£40	£35	£27

R357. AIA-AIA Brush Type 2 Diesel Loco.

Blue. With Magnadhesion. Price 49/6d. Issued 1962. Deleted 1970. 22.7cms.

	£25	£20	£15

R358. 2-6-0 'Davy Crockett' Old Time Loco.

Red, black and yellow. Crew and smoke. complete with 'Davy Crockett' and 'TRR' decals. Price 57/6d. Issued 1963. Deleted 1970. Length 17.1cms. Matching tender is R233. Price 7/6d. 9cms.

	£55	£50	£40

R359. 0-4-0 Tank Loco.

In black livery, this is an electric 12-15 volts DC. Price 27/6d. Issued 1962. Deleted 1970. 10.5cms.

	£12	£9	£7

R386. 4-6-2 'Princess Elizabeth'

Green. With crew and tender. A good investment. Price 52/6d. Issued 1966. Deleted 1973. 17.1cms. approx.

	£40	£35	£25

R550. Saddle Tank Clockwork Loco.

Red and black. Price 27/6d. Issued 1965. Deleted 1970. 12.2cms.

	£20	£15	£10

R553. 4-2-2 'The Caledonian'

Blue and black. Crew and Magnadhesion. Price 47/6d. Issued 1963. Deleted 1970. Matching tender is R554. Loco 13cms. Tender 9.5cms.

	£45	£40	£30

R555. Diesel Pullman Motor Car.

Blue, white and grey. Type 2 loco with seating. Price 55/-. Issued 1965. Deleted 1970. 22cms. approx.

	£20	£15	£10

R556. Diesel Pullman Motor Car. (Non-Power).

Blue, white and grey. Model made to match R555. Type 2 with seating. Price 17/6d. Issued 1965. Deleted 1970. 22cms. approx.

	£12	£9	£7

R559. 0-4-0 Diesel Locomotive.

Green, silver and black. Price 41/-. Issued 1964. Deleted 1970. 10cms.

	£10	£8	£6

MODEL	MB	MU	GC

R644. Bo-Bo Electric Loco.

Blue, yellow and black. Price 75/-. Issued 1967. Deleted 1973.
22cms. approx.

	£15	£11	£9

R645. Hymek Diesel Loco.

Blue, orange, yellow and black. Price 67/6d. Issued 1968. Deleted
1973. 22cms. approx.

	£15	£12	£10

R653. 2-6-2 Continental 'Prairie' Tank Loco.

Black livery with silver and red lines. Rare and unusual loco with
Magnadhesion. Price 52/6d. Issued 1964. Deleted 1968. 17.6cms.

	£55	£50	£40

R751. English Electric Co-Co Diesel.

Green with grey roof. One of the latest Type 3 Diesel locos for
mixed traffic to go into large scale service with British Railways.
Price 75/-. Issued 1965. Deleted 1973. 22cms. approx.

	£20	£15	£10

R752. Battle Space Turbo Car.

Dark red with dark orange nose and propeller. Zig-zag space design
on sides. A very popular model. Variable speed, propeller-driven
armoured space car with heavy duty ramming spike for attack
missions. Price 52/6d. Issued 1968. Deleted 1973. 8.7cms.

	£30	£25	£20

R753. Co-Co E3,000 Class Electric Loco.

Blue, black and silver. Working pantographs and Magnadhesion.
Operating from the high-low dual control system, either from the
track, or, at the flick of a switch, from the overhead power supply
system. Price 65/-. Issued 1964. Deleted 1970. 22.2cms.

	£20	£25	£10

R754. 0-4-4 M7 Class Tank Loco.

Black livery, with '30021' on sides. Fitted with hinged smoke box
door which opens to show boiler tube detail. Crew. The cab has a
realistic glowing firebox and space to stand driver and fireman.
Price 50/9d. Issued 1967. Deleted 1973. 12.1cms. approx.

	£20	£15	£10

R758. Hymek B-B Diesel Hydraulic Loco.

Green, white, black and yellow. Number 3D95 on front plate and
number D7000 on sides. Price 59/3d. Issued 1966. Deleted 1970.
22cms.

	£15	£12	£9

R759. 4-6-0 'Albert Hall' Class Loco.

Authentic green Western Region livery, with Magnadhesion.
Complete with driver and fireman. Price 61/-. Issued 1966. Deleted
1973. 16cms. approx.

	£50	£45	£35

R850. 'The Flying Scotsman'.

Green and black livery, with matching R851 tender. As 1968 was the last year for regular steam services in Great Britain, the Triang Company issued this model as they considered it to be probably the most famous of all steam locomotives. Fortunately, the original Flying Scotsman has been preserved and can be seen at York Railway Museum. Model was fitted with an internal light to simulate the glow from the fire box. It has the driver and fireman, scale diameter driving wheels, corridor tender, as well as Magnadhesion. Price 65/-. Issued 1968. Deleted 1973. A good investment. Some of the first models were in a special presentation box and dated as such. These boxed models are worth considerably more than ordinary later issues and therefore I have listed a separate price. 22cms. approx.

Price for special dated boxed model	£100		
Price for normal model	£45	£40	£35

RW397. The Satellite Train.

An All action train with computer radar control centre and flying satellite. Red, black and blue, although colours could vary. Radar scanner on control car rotates as the train fires as it moves. Price 63/-. Issued 1965. Deleted 1970. 16cms. approx.

	£15	£10	£8

RW398. Strike Force 10 Train.

Red and black, although colours could vary. This special train is equipped for air and ground warfare. Price 79/6d. Issued 1965. Deleted 1970. 22cms. approx.

	£15	£10	£8

RW2207. 0-6-0 Tank Locomotive.

Green and black with B.R. decals. Price 38/-. Issued 1965. Deleted 1970. 12.1cms. approx.

	£15	£12	£10

RW2217. 0-6-2 Tank Locomotive.

Black with red and green lines and B.R. decals. Also number 69550. Price 56/9d. Issued 1965. Deleted 1970. 17.8cms. approx.

	£20	£15	£9

RW2218. 2-6-4 Class 4 MT Tank Loco.

Black with BR emblems and number 80033 on sides. Price 65/-. Issued 1969. Deleted 1973. 13cms. approx.

	£25	£20	£15

RW2221. 4-6-0 'Cardiff Castle' Class 7P Loco.

The first of a series of super-heavy locomotives built for Triang Hornby by Wrenn when the Triang company amalgamated with Hornby officially in 1968. They have fine body detailing, individual hand rails and nickelled tyres on their driving wheels. They operate from 12 volts DC and they are each fitted with a Triang Hornby coupling. Price 75/-. Issued 1968. Deleted 1973, although these models were still supplied by the new Hornby Company when they took over in 1972 at the end of the official Triang run. 16cms. approx. With matching tender 8.7cms.

	£50	£45	£35

RW2224. 2-8-0 Class AF Loco with Tender.

Black and silver livery. Price 69/6d. Issued 1968. Deleted 1973. 26cms. approx.

	£40	£35	£30

RW2226. Duchess Class 'City of London' Loco with Tender.

Maroon and black with City of London decals. One of the better class locos and almost as good an investment as the 'Flying Scotsman'. Price 95/-. Issued 1968. Deleted 1973, although again this loco was carried on by the new Hornby Company. 26cms.

	£75	£65	£50

MODEL	MB	MU	GC

RW2231. 0-6-0 Diesel Electric Shunter.

Two-tone green and black with B.R. decals. Price 64/-. Issued 1965. Deleted 1970. 18cms. approx.

| | £12 | £10 | £8 |

RW2233. Co-Bo Diesel Electric Locomotive.

Green, black and silver. An ex-Hornby-Dublo model which came complete with the Triang Hornby coupling converter wagon. Price 61/-. Issued 1966. Deleted 1973. 18cms. approx.

| | £15 | £10 | £7 |

RW2235. 4-6-2 West Country 'Barnstaple' Class 7P5F Loco.

Green and Black with red line design and Barnstaple emblems and decals. Complete with tender. Price 95/-. Issued 1968. Deleted 1973. Length with tender 26cms. approx.

| | £40 | £35 | £25 |

RW2245. E. 3002 Electric Loco

Blue with black chassis and white roof, complete with pantograph. One of the first models produjced after amalgamation of Triang Railway and Hornby Dublo. Price 75/-. Issued 1965. Deleted 1970. 22cms. approx.

| | £15 | £10 | £8 |

RW2250. Electric Motor Coach.

Green with a powered bogie. Price 64/-. Issued 1965. Deleted 1970. 22cms. approx.

| | £20 | £15 | £10 |

RW4150. Electric Driving Trailer Coach.

Model made to go with RW2250 in matching livery. Price 10/6d. Issued 1965. Deleted 1970. 22cms. approx.

| | £10 | £7 | £5 |

COACHES

MODEL	MB	MU	GC
R20. L.M.S. Coach. Maroon, and black with grey roof. Price 10/6d. Issued 1955. Deleted 1970. 17.1cms. approx.	£7	£5	£3
R21. B.R. Coach. White, maroon and grey. Price 10/6d. Issued 1955. Deleted 1970. 19cms.	£7	£5	£3
R22. S.R. Coach. Green Southern Railway livery with black chassis and grey roof. Price 11/6d. Issued 1955. Deleted 1970. 19cms.	£10	£8	£6
R23. Royal Mail Coach. Maroon, white and grey with black chassis. Price 11/6d. Issued 1955. Deleted 1970. 19cms. approx.	£15	£12	£10
R24. Silver Coach. Silver and black with Triang Railways decals. Price 11/6d. Issued 1955. Deleted 1970. 25.1cms.	£15	£12	£10
R25. Vista Dome Coach. Silver and black with domed seats and Triang Railway decals. Price 12/6d. Issued 1956. Deleted 1970. 25.1cms.	£15	£12	£10
R26. L.N.E.R. Coach. Maroon. Price 10/9d. Issued 1956. Deleted 1970. 14.3cms.	£15	£12	£10
R27. Pullman Coach. Red, black and silver. Complete with seats. Price 12/6d. Issued 1956. Deleted 1970. 22.5cms.	£20	£15	£10
R28. B.R. Main Line Brake Third Coach. Maroon and cream. Grey roof and black chassis. Price 11/9d. Issued 1955. Deleted 1970. 22.7cms.	£12	£10	£8
R29. B.R. Main Line Composite Coach. Maroon, cream, grey and black. Price 11/9d. Issued 1955. Deleted 1970. 22.7cms.	£12	£10	£8
R111. Hopper Car. Green and black with Triang decals. Price 9/6d. Issued 1955. Deleted 1970. 14.4cms.	£10	£8	£6
R114. Box Car. Orange, black and silver grey. Price 9/6d. Issued 1955. Deleted 1970. 14.4cms.	£7	£6	£5
R115. Caboose. Red or maroon, black and silver. With 'Triang Railways 7482' decals in white. Price 10/6d. Issued 1959. Deleted 1970. 14.3cms.	£14	£12	£10
R119. T.C. Mail Coach. Black and silver with Triang Railway decals. Price 12/6d. Issued 1955. Deleted 1970. 25.1cms.	£50	£45	£40
Price for normal maroon and silver	£14	£12	£10

MODEL	MB	MU	GC
R120. B.R. Coach. Maroon, black and silver. Price 10/9d. Issued 1955. Deleted 1970. 22.7cms.	£12	£10	£8
R121. B.R. Suburban Composite Coach. Maroon, black and silver. Price 39/6d. Issued 1959. Deleted 1970. 22.7cms.	£14	£12	£10
R123. Horse-Box. Maroon, black and grey. Price 9/6d. Issued 1964. Deleted 1970. 17cms. approx.	£10	£8	£6
R125. Observation Coach. Silver and black with streamlined finish and 'Triang Railways' decals in black. Seats in front domed area. Price 12/6d. Issued 1956. Deleted 1970. 25.1cms.	£15	£14	£12
R130. Baggage Car. Black and silver grey with blue sliding doors. With 'Triang Railways Baggage Car' decals. Price 12/6d. Issued 1957. Deleted 1970. 25.1cms.	£15	£14	£12
R131. Coach. Blue and yellow with 'Triang Railway' decals. Price 12/6d. Issued 1960. Deleted 1970. 25.1cms.	£15	£13	£10
R132. Vista Dome Car. Blue and yellow. Price 12/6d. Issued 1960. Deleted 1970. 25.1cms.	£15	£13	£10
R133. Observation Car. Blue and yellow with 'Triang Railway' decals. Price 12/6d. Issued 1960. Deleted 1970. 25.1cms.	£15	£13	£10
R134. Baggage Car. Blue and yellow with 'Triang Railway' decals. Price 12/6d. Issued 1960. Deleted 1970. 25.1cms.	£15	£13	£10
R220. S.R. Main Line Brake 3rd Coach. Green and grey. Price 12/6d. Issued 1957. Deleted 1970. 22.7cms.	£15	£12	£10
R221. S.R. Main Line Composite Coach. Green and grey. Price 12/6d. Issued 1957. Deleted 1970. 22.7cms.	£15	£12	£10
R222. S.R. Suburban Brake 3rd Coach. Green and grey. Price 12/6d. Issued 1957. Deleted 1970. 22.7cms.	£15	£12	£10
R223. S.R. Suburban Composite Coach. Green and grey. Price 12/6d. Issued 1957. Deleted 1970. 22.7cms.	£15	£12	£10
R224. B.R. Resaurant Car. B.R. maroon, cream and grey with restaurant car decals. Price 14/6d. Issued 1957. Deleted 1970. 22.7cms.	£15	£12	£10
R225. B.R. Main Line Coach. Maroon and cream. Price 12/6d. Issued 1960. Deleted 1970. 22.7cms.	£14	£12	£10
R226. Utility Van. Green and grey. Price 10/6d. Issued 1961. Deleted 1970. 22.7cms.	£12	£10	£8

R226B. Utility Van.

Blue with grey roof and twelve opening doors. Price 14/6d. Issued
1968. Deleted 1973. 22.7cms.

£12 £10 £8

R227. Utility Van.

Maroon and grey with twelve opening doors. Price 12/6d. Issued
1961. Deleted 1970. 22.7cms.

£12 £10 £8

R228. Pullman First Class Car.

Dark brown, Cream and grey. With names which could vary
according to name boards provided e.g. 'Anne'. Price 12/6d. Issued
1960. Deleted 1970. 22.5cms.

£18 £15 £12

R230. Coach.

Red, white and silver grey. Price 11/6d. Issued 1961. Deleted 1970.
18.7cms.

£12 £10 £8

R231. Coach.

Green, white and silver grey. Price 11/6d. Issued 1961. Deleted
1970. 18.7cms.

£12 £10 £8

R248. Ambulance Car.

Green, white and silver grey with Red Cross decals on roof and
sides. With 'R.A.M.C. Ambulance' decals on sides. Price 12/6d.
Issued 1964. Deleted 1970. 25.2cms.

£18 £15 £12

R320. B.R. Main Line Brake 2nd Coach.

B.R. maroon livery. Price 11/6d. Issued 1961. Deleted 1970.
22.7cms.

£12 £10 £8

R321. B.R. Main Line Composite Coach.

B.R. maroon livery. Price 11/6d. Issued 1961. Deleted 1970.
22.7cms.

£12 £10 £8

R322. B.R. Restaurant Car.

B.R. maroon livery. Price 12/6d. Issued 1961. Deleted 1970.
22.7cms.

£15 £13 £10

R324. Diner.

Red and silver with 'Triang Railway' decals. Price 12/6d. Issued
1961. Deleted 1970. 25.1cms.

£15 £12 £10

R325. Diner.

Blue and yellow. Price 12/6d. Issued 1961. Deleted 1970. 25.1cms.

£15 £12 £10

MODEL	MB	MU	GC

R328. Pullman Brake 2nd Car.
Chocolate and cream with grey roof. Price 12/6d. Issued 1960.
Deleted 1970. 22.5cms.

£15 £12 £10

R329. W.R. Main Line Brake 2nd Coach.
Brown, lemon and white, authentic Western Railway colours. Price
11/6d. Issued 1960. Deleted 1970. 22.7cms.

£10 £8 £6

R330. W.R. Main Line Composite Coach.
Brown, lemon and white, authentic Western Railway colours. Price
11/6d. Issued 1960. Deleted 1970. 22.7cms.

£10 £8 £6

R331. W.R. Restaurant Car.
Brown, lemon and white, authentic Western Railway colours. Price
12/6d. Issued 1960. Deleted 1970. 22.7cms.

£10 £8 £6

R332. G.W.R. Composite Coach.
G.W.R. livery of chocolate and cream with a clerestory roof. Price
12/6d. Issued 1961. Deleted 1970. 19.1cms.

£18 £15 £12

R333. G.W.R. Brake 3rd Coach.
G.W.R. livery of chocolate and cream with a clerestory roof. Price
12/6d. Issued 1961. Deleted 1970. 19.1cms.

£18 £15 £12

R334. Diesel Rail Car Centre Unit.
Green and grey livery for use in connection with R157 and R158
diesel rail cars. Price 12/6d. Issued 1961. Deleted 1970. 21.7cms.

£15 £12 £10

R335. Coach.
Green two-tone and grey with Triang Railway decals. Price 12/6d.
Issued 1960. Deleted 1970. 25.2cms.

£15 £12 £10

R336. Observation Car.
Green two-tone and grey with 'Triang Railway' decals. Price 12/6d.
Issued 1961. Deleted 1970. 25.2cms.

£15 £12 £10

R337. Baggage/Kitchen Car.
In green two-tone and grey with 'Triang Railway' decals. Price
12/6d. Issued 1961. Deleted 1970. 25.2cms.

£15 £12 £10

R338. Dining Car.
Green two-tone and grey with 'Triang Railway' decals. Price 12/6d.
Issued 1961. Deleted 1970. 25.2cms.

£15 £12 £10

R339. B.R. 2nd Class Sleeping Car.
B.R. maroon with grey roof and sleeping car decals. Sleeping cars
have eleven compartments, each with two berths and a hand basin.
Full bedding is provided and an attendant's compartment is
equipped for supplying tea and biscuits. Price 13/6d. Issued 1961.
Deleted 1970. 26.2cms.

£18 £15 £12

R382. C.K.D. Composite Coach.
Maroon. Price 18/11d. Issued 1965. Deleted 1970. Boxed in pairs.
Each coach 26.2cms. approx.
Price for single coach

£12 £10 £8

Price for pair

£25

MODEL	MB	MU	GC
R383. C.K.D. Brake 2nd Coach.			
Maroon. Price 18/11d. Issued 1965. Deleted 1969. Boxed in pairs. Each coach 26.2cms. approx.			
Price for single coach	£12	£10	£8
Price for pair	£25		
R400. Transcontinental Mail Coach.			
Red with grey roof and 'Transcontinental' decals. Price 12/6d. Issued 1962. Deleted 1970. 25.1cms.	£15	£12	£10
R401. Transcontinental Mail Coach.			
Blue with yellow lines and 'Transcontinental' decals. Price 12/6d. Issued 1962. Deleted 1970. 25.1cms.	£18	£15	£12
R402. Royal Mail Coach.			
Dark maroon or medium red. The latter being worth double. With Royal Mail decals. Grey roof. Price 11/6d. Issued 1962. Deleted 1973. 19.4cms.	£12	£10	£8
R422. B.R. 1st/2nd Composite Coach.			
Maroon and with built-in seat unit with grey roof. Price 12/6d. Issued 1962. Deleted 1970. 26.2cms.	£14	£12	£10
R423. B.R. Brake 2nd Coach.			
Maroon with grey roof and built-in seat unit. Price 12/6d. Issued 1962. Deleted 1970. 26.2cms.	£14	£12	£10
R424. Buffet Car.			
Maroon with grey roof and built in interior fittings. Price 12/6d. Issued 1962. Deleted 1970. 26.2cms.	£16	£14	£12
R425. B.R. Full Parcels Brake Coach.			
Maroon with grey roof and built in interior fittings. Price 12/6d. Issued 1962. Deleted 1970. 26.2cms.	£16	£14	£12
R426. Pullman Parlour Car Type 6.			
Blue, grey and white with interior seats. Very rare coach made in a limited number. Price 13/6d. Issued 1964. Deleted 1970. 26.4cms.	£20	£15	£10
R427. Caledonian 1st/3rd Composite Coach.			
Green and grey. Rare model. Price 12/6d. Issued 1964. Deleted 1970. 26.2cms.	£50	£40	£30
R428. Caledonian Brake/Composite Coach.			
Two-tone maroon, white and black with silver grey roof. Price 12/6d. Issued 1963. Deleted 1970. 26.2cms.	£20	£16	£12
R429. Caledonian 1st/3rd Composite Coach.			
Two-tone maroon, white and black with silver grey roof. Price 12/6d. Issued 1963. Deleted 1970. 26.2cms.	£20	£16	£12
R440. Coach.			
Transcontinental red and silver with built-in seat unit. Complete with 'Transcontinental' decals in black. Price 12/6d. Issued 1962. Deleted 1970. 25.2cms.	£20	£16	£12

MODEL	MB	MU	GC

R441. Observation Car.

Extended dome roof with a streamlined appearance. Red and silver Transcontinental livery with built-in seat unit. Price 14/6d. Issued 1962. Deleted 1970. 25.2cms.

£25 £20 £15

R442. Baggage/Kitchen Car.

Red and silver Transcontinental livery. Price 12/6d. Issued 1962. Deleted 1970. 25.2cms.

£20 £16 £12

R443. Transcontinental Diner.

Red and silver with built-in interior fittings. Price 12/6d. Issued 1962. Deleted 1970. 25.2cms.

£20 £16 £14

R444. Continental Coach.

Blue and yellow with built-in seat unit. Price 12/6d. Issued 1962. Deleted 1970. 25.2cms.

£16 £14 £12

R444CN. Canadian National Passenger Car.

Silver and black. Price 17/6d. A rare item issued in a limited edition in 1967. Deleted 1970. 25.2cms. approx.

£30 £25 £20

R445. Observation Car.

Blue and yellow with built-in seat unit. Price 12/6d. Issued 1962. Deleted 1970. 25.2cms.

£16 £14 £12

R445CN. Canadian National Observation Car.

Silver and black. Price 17/6d. A rare item issued in a limited edition in 1967. Deleted 1970. 25.2cms.

£30 £25 £20

R446. Baggage Kitchen Car.

Blue and yellow livery with 'Transcontinental' decals. Price 12/6d. Issued 1962. Deleted 1970. 25.2cms.

£16 £14 £12

R447. Diner. Transcontinental.

Blue and yellow with built-in interior fittings and 'Transcontinental' decals. Price 12/6d. Issued 1962. Deleted 1970. 25.2cms.

£16 £14 £12

R448. Old Tyme Coach.

Golden orange and chocolate with clerestory roof and seats. Price 10/6d. Issued 1962. Deleted 1970. 19.7cms.

£20 £15 £12

R620. Engineering Department Coach.

Chocolate brown with grey roof. Price 8/11d. Issued 1963. Deleted 1970. 19.1cms.

£30 £25 £20

MODEL	MB	MU	GC
R621. Liverpool/Manchester Coach.			
Yellow, brown and grey, which was Stephenson's original livery for the famous 'Rocket' loco. Price 7/6d. Issued 1963. Deleted 1970. 8.7cms.	£30	£25	£20
R622. S.R. Main Line Composite Coach.			
Green with orange line and silver grey roof. Price 10/6d. Issued 1963. Deleted 1970. 26.2cms.	£20	£16	£12
R623. Main Line S.R. Brake 2nd Coach.			
Green with orange line and silver grey roof. Price 10/6d. Issued 1963. Deleted 1970. 26.2cms.	£20	£16	£12
R624. S.R. Buffet Car.			
Green with orange line and silver grey roof. Price 10/6d. Issued 1963. Deleted 1970. 26.2cms.	£20	£16	£12
R625. Continental Wagon-Lits Sleeping Car.			
Royal blue or purple with golden lines and silver grey roof, complete with built-in interior fittings. With continental and sleeping car decals. Price 11/6d. Issued 1963. Deleted 1970. 11.6cms.	£30	£25	£20
R626. Main Line Composite Coach.			
Crimson and cream with grey roof and 'Main Line' decals. With seats. Price 11/6d. Issued 1963. Deleted 1970. 26.2cms.	£16	£14	£12
R627. Main Line Brake 2nd Coach.			
Crimson and cream with grey roof and 'Main Line' decals. With seats. Price 11/6d. Issued 1963. Deleted 1970. 26.2cms.	£16	£14	£12
R628. Main Line Buffet Car.			
Crimson and cream with grey roof and 'Main Line' decals. With seats. Price 11/6d. Issued 1963. Deleted 1970. 26.2cms.	£16	£14	£12
R722. Inter-City 2nd Coach.			
Blue and white with grey roof and 'Inter-City' decals. Complete with seating. Price 17/6d. Issued 1968. Deleted 1970. 26.2cms.	£15	£13	£10
R723. Inter-City Brake 1st Coach.			
Blue and white with grey roof and 'Inter-City' decals. Complete with seating. Price 17/6d. Issued 1968. Deleted 1973. 26.2cms.	£15	£13	£10
R724. Inter-City 2nd Coach.			
Blue and white with 'Inter-City' decals and detailed interior. A fine model of B.R. Mark 2 Carriage, 2nd Class open. Price 13/6d. Issued 1969. Deleted 1973. 26.2cms.	£15	£13	£10
R725. Diner Car.			
Blue and white with 'Inter-City' decals, and detailed interior. Price 13/6d. Issued 1969. Deleted 1973. 26.2cms.	£15	£13	£10
R726. Inter-City Brake 2nd Coach.			
Blue and white with 'Inter-City' decals and detailed interior. A fine model of B.R. Mark 2 Carriage, 2nd Class open. Price 13/6d. Issued 1969. Deleted 1973. 26.2cms.	£15	£13	£10

MODEL	MB	MU	GC
R727. Blue Composite Coach.			
Blue and white with interior fittings. Coach was recoloured in line with British Rail's new stock for the Euston/Crewe/Manchester/ Liverpool electrification project. Price 13/6d. Issued 1966. Deleted 1970. 26.2cms.	£15	£13	£10
R728. Blue Brake 2nd Coach.			
Blue and white with interior fittings. Coach was recoloured in line with British Rail's new stock for the Euston/Crewe/Manchester/ Liverpool electrification project. Price 13/6d. Issued 1966. Deleted 1970. 26.2cms.	£15	£13	£10
R729. Blue Buffet Car.			
Blue and white with interior fittings. Coach was recoloured in line with British Rail's new stock for the Euston/Crewe/Manchester/ Liverpool electrification project. Price 13/6d. Issued 1966. Deleted 1970. 26.2cms.	£15	£13	£10
R730. Composite Coach.			
An excellent assembly pack and now a very good investment. Blue livery in a special presentation box. Price 13/1d. Issued 1969. Deleted 1973. 26.2cms. approx.	£25	£20	£15
R731. Blue Brake 2nd Coach.			
An excellent assembly pack and now a very good investment. Blue livery in a special presentation box. Price 13/1d. Issued 1969. Deleted 1973. Length when complete 26.2cms. approx.	£25	£20	£15
R732. Blue Buffet Car.			
An excellent assembly pack and now a very good investment. In blue livery in a special presentation box. Price 14/-. Issued 1969. Deleted 1973. Length when complete 26.2cms.	£25	£20	£15

VANS AND WAGONS

MODEL	MB	MU	GC
R10. Goods Truck.			
Grey with 'G.W.' decals. Later green without decal. Price 3/6d. Issued 1955. Deleted 1970. 7.5cms.	£5	£4	£3
R11. Goods Van.			
Red rust with grey or white roof and 'G.W.' decals. Later green. Price 3/11d. Issued 1955. Deleted 1970. 7.5cms.	£6	£5	£4
R11/A.B.R. Ventilated Van.			
Chocolate and grey. Price 8/6d. Issued 1968. Deleted 1970. 7.5cms.	£5	£4	£2
R12. Shell Tank Wagon.			
Silver with B.P. and black star decals on sides. Price 4/-. Issued 1955. Deleted 1970. 7.5cms.	£7	£6	£5
R13. Coal Truck.			
Red rust with imitation coal and 'N.E.' decals. Price 3/11d. Issued 1955. Deleted 1970. 7.5cms.	£5	£4	£3
R14. Fish Van.			
White and grey. Later ice blue with 'Insulfish' on side. Price 3/11d. Issued 1955. Deleted 1970. 7.5cms.	£6	£5	£4
R15. Milk Tank Wagon.			
White with 'U.D.' decals. Price 3/11d. Issued 1955. Deleted 1970. 7.5cms.	£6	£5	£4
R16. Brake Van.			
White or off-white. Later red-rust. Price 3/11d. Issued 1955. Deleted 1970. 7.5cms.	£4	£3	£2
R17. Bolster Wagon.			
Grey or off-white. Price 3/11d. Issued 1955. Deleted 1970. 7.5cms.	£3	£2	£1
R17/C. Flat Wagon with Minic Car.			
Chocolate or brown, complete with a green Minic car, although colours of these can vary. Price 5/1d. Issued 1966. Deleted 1970. 7.5cms.	£9	£7	£5
R18. Cable Drum Wagon.			
Grey or brown, complete with two black cable drums with 'Pirelli X General' decals. One red roll of wire cable and one green roll of wire cable. Price 4/11d. Issued 1955. Deleted 1970.	£5	£4	£3
Drums with green edges. Issued 1964, otherwise as above	£4	£3	£2
R19. Flat Wagon with Tarpaulined Load.			
Grey or brown with grey or green tarpaulined load. Price 5/6d. Issued 1959. Deleted 1970. 16.8cms.	£7	£6	£5
R110. Bogie Bolster Wagon.			
Grey. Price 7/6d. Issued 1955. Deleted 1970. 16.8cms.	£7	£6	£5
Brown	£6	£5	£4
R111. Hopper Car.			
Green with 'Triang' decals. Price 9/6d. Issued 1955. Deleted 1970. 14.4cms.	£10	£7	£5

R112. Goods Truck with Drop Doors.
Light or medium brown. Price 4/6d. Issued 1955. Deleted 1970.
7.5cms.

£6 · £5 · £4

R113. Goods Truck with Drop Sides.
Light or dark blue grey. Later red-rust. Price 4/6d. Issued 1955.
Deleted 1970. 7.5cms.

£6 · £5 · £4

R114. Box Car.
Orange and grey with sliding door and 'Triang Railway' decals.
Price 9/6d. Issued 1955. Deleted 1970. 14.4cms.

£7 · £6 · £5

R115. Caboose.
Red or maroon, black and silver. Complete with 'Triang Railways
7482' decals in white. Price 10/6d. Issued 1955. Deleted 1970.
14.3cms.

£14 · £12 · £10

R116. Gondola.
Grey or dark brown. Latter worth double. With 'Triang Railways
TR3576' decals. Price 7/6d. Issued 1955. Deleted 1970. 14.6cms.

£8 · £7 · £5

R117. Oil Tanker.
Red and black with 'Shell Capacity 100. 1000 lbs.' decals. Price
9/6d. Issued 1955. Deleted 1970. 15.3cms.

£14 · £12 · £10

R117CN. Oil Tanker Car.
Black with 'CN137000' decals. Made especially for the Canadian
National Railways. Price 12/6d. Issued 1968. Deleted 1970.
15.3cms.

£20 · £18 · £15

R118. Bogie Well Wagon.
Orange. Later green. Price 5/11d. Issued 1955. Deleted 1970.
19.2cms.

£15 · £14 · £12

R118/A. Bogie Well Wagon.
Green. Price 5/11d. Issued 1964. Deleted 1970. 19.2cms.

£12 · £10 · £8

R119. Flat Cart.
Grey, brown or green. Price 4/6d. Issued 1956. Deleted 1970.
8cms.

£15 · £12 · £10

R122. Cattle Wagon.
Light or medium brown with long wheel base. Price 5/6d. Issued
1956. Deleted 1970. 10cms.

£9 · £7 · £5

R123. Horse Box.
Dark or medium grey with long wheel base. Price 5/6d. Issued
1956. Deleted 1970. 10cms.
In Brown. Price 39/6d. Issued 1959. Deleted 1970. 22.7cms.

£9 · £7 · £5
£7 · £6 · £4

R124. W.R. Brake Van.
Brown livery with long wheel base. Price 5/11d. Issued 1956.
Deleted 1970. 10cms.

£9 · £7 · £5

R126. Stock Car.
Yellow and black with sliding doors. Price 8/11d. Issued 1960.
Deleted 1970. 18.3cms.

£15 · £12 · £10

R127. Operating Crane Truck.

Brown and black with lifting, rising and rotating gib. Price 14/6d.
Issued 1963. Deleted 1970. 9.7cms.   £15   £12   £10

R128. Operating Helicopter Car.

Green or dark brown. Complete with either red or yellow
helicopter (R165). Latter worth double. Price 14/6d. Issued 1964.
Deleted 1970. 18.3cms. Helicopter 10.8cms.   £20   £18   £15

R129. Refrigerator Car.

White and grey with 'T.R. Refrigerator TR2690' decals in blue and
red on sides. Price 8/6d. Issued 1956. Deleted 1970. 18.3cms.   £15   £12   £10

R136. Box Car.

Grey and black with sliding doors and 'T.R. Speedy Arrow Service'
decals in red and yellow on sides. Price 9/6d. Issued 1960. Deleted
1970. 18.3cms.   £15   £12   £10

R137. Cement Car.

Grey and black with a gravity unloading mechanism. Price 11/6d.
Issued 1960. Deleted 1970. 14.3cms.   £15   £12   £10

R138. Snow Plough

Green and blue grey with opening side wings. Price 10/6d. Issued
1963. Deleted 1970. 15.4cms.   £20   £16   £15

R139. Pickle Car.

Grey and light green with 'Westwood Pickles' decals. Complete
with four pickle containers. Price 11/6d. Issued 1960. Deleted 1970.
15.4cms.   £20   £15   £14

MODEL	MB	MU	GC
R210. Shell Fuel Oil Tank Wagon.			
Black 'Shell BP' decals in white. Complete with Star. Price 5/6d. Issued 1957. Deleted 1970. 7.5cms.	£10	£8	£6
R211. Shell Lubricating Oil Tank Wagon.			
Gold and black with 'Shell Lubricating Oil' in red and a black star decal. Price 5/6d. Issued 1957. Deleted 1970. 7.5cms.	£10	£8	£6
R212. Bogie Bolster Wagon.			
Grey with brown log load. Price 5/11d. Issued 1957. Deleted 1970. 16.8cms.	£12	£10	£8
R213. Bogie Well Wagon with Crane Load.			
Orange wagon with blue crane with black wheels. Price 7/6d. Issued 1957. Deleted 1970. 19.2cms.	£20	£16	£15
R214. Ore Wagon.			
Dark lime green with unloading gravity mechanism. Price 5/6d. Issued 1961. Deleted 1970. 10cms.	£7	£6	£5
R214/A. Hopper Wagon.			
Lime green. Price 7/6d. Issued 1969. Deleted 1970. 10cms. approx.	£7	£6	£5
R215. Bulk Grain Wagon.			
Lime green with gravity unloading mechanism. Price 5/6d. Issued 1961. Deleted 1970. 10cms.	£7	£6	£5
R216. Rocket Launching Wagon.			
Grey wagon with yellow and grey rocket launcher complete with red and silver rocket. Price 12/6d. Issued 1961. Deleted 1970. 16.8cms.	£18	£16	£14
R217. Open Truck.			
Dull lime green. Price 5/11d. Issued 1961. Deleted 1970. 9.4cms.	£7	£6	£5
R218. Closed Van.			
Rich dark green with white roof. Price 5/11d. Issued 1961. Deleted 1970. 9.4cms.	£10	£8	£6
R219. Bogie Brick Wagon.			
Brick red and black with brick decals in white on sides. Price 6/11d. Issued 1962. Deleted 1970. 16.8cms.	£12	£10	£8
R234. Flat Car.			
Slate grey. Price 6/11d. Issued 1961. Deleted 1970. 18.3cms.	£7	£6	£5
R235. Pulp Wood Car.			
Slate grey with brown pulp wood load. Price 9/6d. Issued 1961. Deleted 1970. 18.3cms.	£12	£10	£8
R236. Depressed Centre Car.			
Lime green. Price 6/11d. Issued 1961. Deleted 1970. 18cms.	£9	£7	£5
R237. Depressed Centre Car with Low Loader Bulldozer Load.			
Lime green car with red, yellow and white bulldozer low-loader load. Price 12/6d. Issued 1961. Deleted 1970. 18cms.	£18	£15	£12

R238. Depressed Centre Car with Transatlantic Cable Drums Load.

	MB	MU	GC
Blue car with black and lemon cable drum load. 'British Insulated Cables' decals in white. Price 11/6d. Issued 1960. Deleted 1970. 18cms.	£12	£10	£8

R239. Bomb Transporter.

	MB	MU	GC
Lime green transporter with red and silver bomb load with 'Triang Minic' decals. The removable bomb load can be charged with caps, and when hand launched it goes off on hitting a hard surface. Price 12/6d. Issued 1964. Deleted 1970. 26.8cms.	£20	£16	£14

R239K. Red Arrow Bomb Transporter.

	MB	MU	GC
Grey transporter, with deep maroon and silver capped bomb. Price 15/6d. Issued 1966. Deleted 1970. 26.8cms.	£20	£16	£14

R240. Bogie Brick Wagon.

	MB	MU	GC
Rust red with brick load. Price 7/6d. Issued 1960. Deleted 1970. 16.8cms.	£10	£8	£6

R241. Bogie Well Wagon with Conqueror Tank.

	MB	MU	GC
Wagon in grey and tank in military green. Price 12/6d. Issued 1961. Deleted 1970. 19.2cms.	£20	£16	£14

R242. Trestrol Wagon.

	MB	MU	GC
Grey. Price 7/11d. Issued 1961. Deleted 1970. 26.8cms.	£12	£10	£8

R243. Mineral Wagon.

	MB	MU	GC
Dull green. Price 4/6d. Issued 1961. Deleted 1970. 7.5cms.	£7	£6	£5

R244. Mineral Wagon with Coal Load.

	MB	MU	GC
Dull green with black coal load. Price 5/6d. Issued 1961. Deleted 1970. 7.5cms.	£8	£7	£6

R245. Open Wagon with Oil Drum Load.

	MB	MU	GC
Grey wagon with fawn load. Price 5/6d. Issued 1961. Deleted 1970. 7.5cms.	£8	£7	£6

R246. Open Wagon with Timber Load.

	MB	MU	GC
Grey or green with fawn timber load. Price 5/11d. Issued 1959. Deleted 1970. 7.5cms.	£8	£7	£6

R247. Bogie Tank Wagon.

	MB	MU	GC
Bright orange with black ladder and chassis. Complete with I.C.I. decals. Price 10/6d. Issued 1962. Deleted 1970. 13cms.	£18	£16	£14

MODEL	MB	MU	GC

R248. Sugar Container Wagon.
Blue with 'Tate and Lyle' decals in white. Rare model. Price 12/6d.
Issued 1961. Deleted 1970. 13cms.

	£30	£25	£20

R249. Exploding Car.
Bright orange with various decals such as 'Danger', 'Warheads',
'Handle Carefully' etc. One only has to hit this lethally laden
freight car with a missile from either of the Triang Railways rocket
launchers and it will explode most realistically. Easy to reassemble
again and again. Price 12/6d. Issued 1963. Deleted 1970. 14.5cms.

	£16	£14	£12

R250. Rank Flour Mills Wagon.
White and black with gold lettering. Another rare item made in a
limited supply. Price 10/6d. Issued 1963. Deleted 1966. 13cms.

	£50	£45	£40

R262. Continental Guards Van.
Dark brown or chocolate. Price 6/6d. Issued 1963. Deleted 1970.
9cms.

	£15	£12	£10

R340. Three Containers Wagon.
Rust red and black with three green containers. Price 4/6d. Issued
1963. Deleted 1970. 7.5cms.

	£6	£5	£4

R341. Searchlight Wagon.
Green and grey with red and blue chevrons and working silver and
red searchlight. Operates from 12 volts DC picking up electric
currents from the track. Price 21/-. Issued 1964. Deleted 1970.
26.8cms.

	£25	£20	£15

R346. Car Transporter.
Pink, complete with six Minix cars, each having windows, plated
grille, bumpers and chassis. Cars normally provided are golden
yellow, royal blue, medium green, bright red and silver. Silver is
very rare. Price 13/9d. Issued 1965. Deleted 1970. 26.8cms.

	£35	£30	£25

R343. Rocket Launcher.
Green and grey with blue and red chevrons. Complete with four
rockets which can be launched by hand simultaneously or
independently. Firing tower rotates to line up projectiles. Price
24/-. Issued 1964. Deleted 1970. 26.8cms.

	£30	£25	£20

R344. Track Cleaning Car.
Black with silver blue roof. With six charges of track cleaning fluid
and spare felt cleaning pad. Price 9/6d. Issued 1964. Deleted 1970.
14.3cms.

	£15	£12	£10

R345. Side Tipping Flat Car.
Orange, blue and black with track side operating ramp and
collection bin which clip fits to the track. As flat car passes
collection bin, the carrying frame tilts and logs or other baggage
rolls off into bin. Price 15/6d. Issued 1964. Deleted 1970. 14.7cms.

	£15	£14	£10

R346. Crew Repair Wagon.
Green, grey or black. Price 7/6d. Issued 1965. Deleted 1970.
10cms.

	£25	£20	£15

R347. Engineering Department Wagon.
Lime green and black with gravity unloading mechanism. Price
6/6d. Issued 1963. Deleted 1970. 10cms.

	£10	£7	£5

MODEL	MB	MU	GC

R348. Giraffe Car.

Chocolate and blue with line side tell tale which clip fits to track.
Also in yellow and orange, worth double. Giraffe is yellow and
black, though colours may vary. Price 21/-. Issued 1964. Deleted
1970. 16.5cms.

	£15	£12	£10

R349. Bogie Chlorine Tank Wagon.

Black or dark green. With 'Murgatroyd's Liquid Chlorine' decals.
Price 10/6d. Issued 1963. Deleted 1970. 13cms.

As above	£50	£45	£40
Off-white or cream with black decals.	£25	£20	£15

R449. Old Time Caboose.

Rust red and grey. Price 7/6d. Issued 1964. Deleted 1970. 9cms.

	£20	£15	£10

R475. Platform Truck with Operating Crane.

Grey truck with brown crane. Price 9/6d. Issued 1964 Deleted
1970. 18cms.

	£20	£15	£10

R560. Transcontinental Crane Car.

Royal green car with rust red working crane. Price 14/6d. Issued
1963. Deleted 1970. 18cms.

	£25	£20	£15

R561. Triang Container Wagon.

Orange red bogie with blue container load with 'Triang Pedigree
Toys and Prams' decals in red and white. Price 5/6d. Issued 1963.
Deleted 1970. 7.5cms.

	£15	£12	£10

R562. Catapult Plane Launch Car.

Grey and lime green livery with orange and blue plane with 'RAF'
decals. The close support fighter plane flies over 15ft. when
launched by trackside trigger. Price 17/6d. Issued 1969. Deleted
1970. 26.8cms.

	£25	£20	£15

R563. Bogie Bolster Wagon.

Black and fawn bogie with three Minix Ford vans in red, white and
blue. Price 11/3d. Issued 1967. Deleted 1970. 13cms. approx.

	£20	£16	£14

R564. Cement Wagon.

Silver grey with 'Blue Circle' cement decals. Price 6/11d. Issued
1966. Deleted 1970. 7.5cms.

	£10	£8	£6

R571. G-10 'Q' Car.

Dark military green with large G-10 decals in yellow on sides. A
plain freight wagon which houses two Red-eye rocket launchers. As
wagon passes track trigger, roof and sides fall away and rocket
launchers swing round to 90° and elevate into firing position.
Complete with two black and white rocket launchers. Price 14/6d.
Issued 1967. Deleted 1970. 13cms. approx.

	£20	£16	£14

R577. Converter Wagon.

Grey and black this is a Triang Hornby converter wagon with a
Triang coupling at one end and a Hornby Dublo coupling at the
other, so wagon can be used in trains of mixed stock. Complete
with a Triang Hornby tension lock coupling. Price 2/6d. Issued
1966. Deleted 1970. 8cms. approx.

	£7	£6	£5

MODEL	MB	MU	GC
R630. P.O.W. Car Military green. This transporter was used to carry prisoners of war away from the combat zone. It also carried freight and ammunition. Price 12/10d. Issued 1966. Deleted 1970. 19.7cms. approx.	£15	£14	£12
R631. Tank Recovery Wagon. Military green with 901 decals in yellow. Rail mounted heavy-duty crane for vehicle recovery and unloading heavy supplies. Price 17/11d. Issued 1967. Deleted 1970. 25.2cms. approx.	£10	£8	£6
R632. Armoured Rail Car. Military green. Used for fighting in areas where the stretches of rail ran for several miles, especially in war zones. Price 15/6d. Issued 1967. Deleted 1970. 19.7cms. approx.	£30	£25	£20
R633. Liner Train. Black wagon with tree red and grey containers. 'Freight Liner' decals on large red band. Each container has opening doors at one end. Price 15/6d. Issued 1967. Deleted 1970. 25.2cms. approx.	£12	£10	£8
R636. Guards Van. Brown, black and grey. Price 9/6d. Issued 1967. Deleted 1970. 10cms.	£15	£12	£10
R639. Sniper Car. Military green with orange and black decals. A hidden sniper in grey suddenly springs up to fire when a trigger catch is released. Price 12/10d. Issued 1967. Deleted 1970. 19.7cms. approx.	£15	£14	£12
R647. Dewars Bulk Grain Wagon. Blue with 'Dewars' decals. Price 10/6d. Issued 1968. Deleted 1970. 10cms.	£12	£10	£8
R648. Johnny Walker Bulk Grain Wagon. Blue with 'Good Old Johnny Walker' decals in black and white. Price 10/6d. Issued 1968. Deleted 1970. 10cms.	£12	£10	£8
R649. VAT 69 Bulk Grain Wagon. Blue with 'VAT 69' decals in white. Price 10/6d. Issued 1968. Deleted 1970. 10cms.	£12	£10	£8
R650. Haig Bulk Grain Wagon. Blue with 'Haig Whisky' decals. Price 10/6d. Issued 1968. Deleted 1970. 10cms.	£12	£10	£8
R666. B.R. Cartic Car Carrier. Green or grey with 16 multi-coloured scale model Minic Cars. Cars in orange, yellow, green, purple, blue, and rose pink although colours can vary. Price 21/-. Issued 1969. Deleted 1970. 25.8cms.	£45	£40	£30
R668. Bowaters 'China Clay' Slurry Wagon. Sky blue with 'Bowater' decals in white. Price 6/6d. Issued 1969. Deleted 1970. 7.5cms.	£7	£6	£5
R725. Command Car. Military green with 'Command Car' decals. This car was used for Battle Group Headquarters for Staff Officers. It picks up and delivers despatches, maps, rations and ammunition. Price 17/11d. Issued 1969. Deleted 1970. 25.8cms. approx.	£25	£20	£15

BUILDINGS, SIGNALS, CROSSING AND ACCESSORIES

MODEL	MB	MU	GC
R43D. Distant Signal. Black, yellow and white. Hand operated distant signal. Price 3/11d. Issued 1955. Deleted 1970. Height 10.8cms.	£5	£4	£3
R43H. Home Signal. Black, white and red. Hand-operated home signal. Price 3/11d. Issued 1955. Deleted 1970. Height 10.8cms.	£5	£4	£3
R45. Turntable Set. Grey and black. Electrically operated from 12 volts DC by remote control. Price 55/-. Issued 1955. Deleted 1970.	£10	£7	£5
R60. Ticket Office. Red, yellow, grey and black. Price 12/6d. Issued 1955. Deleted 1970. 12.7cms.	£12	£10	£8
R61. Signal Box. Orange, red, grey and yellow with blue sign. Price 3/11d. Issued 1955. Deleted 1970. 9cms.	£9	£7	£5
R62. Waiting Room. Red, grey, orange and yellow. Price 7/6d. Issued 1955. Deleted 1970. 9.2cms.	£10	£8	£6
R63. Central Platform Unit. White and grey. Price 1/11d. Issued 1955. Deleted 1970. 18.1cms.	£3	£2	£1
R64. Platform Curved End. Grey and white. These curved ends were made in both left and right. Price 1/11d. Issued 1955. Deleted 1970. 17.9cms.	£3	£2	£1
R65. Platform Short Ramp. Grey and white. Price 1/11d. Issued 1965. Deleted 1970. 4.4cms.	£3	£2	£1
R66. Porters Room. Red, yellow, grey and black, complete with 'Travel Triang Railways' sign on side in pink, yellow and white. Price 3/11d. Issued 1955. Deleted 1970. 9.5cms.	£5	£4	£3
R66K. Porters Room with Kiosk. Red, yellow, grey, black and green with 'Margate' sign on side, in red, blue, yellow and black. Neatly set out kiosk and adverts etc. Price 4/6d. Issued 1955. Deleted 1970. 9.5cms.	£9	£7	£5
R67. Approach Steps. Grey and white with 'We Want Watney's Brown Ale', plus airport poster. Price 1/11d. Issued 1955. Deleted 1970. 9.2cms.	£4	£3	£2
R68. Station Name Board. Yellow with blue sign and 'Chatham' decal. Name board and fence section complete. Price 1/11d. Issued 1955. Deleted 1970. Name board 8.9cms. Fence section 4.4cms.	£3	£2	£1
R69. Newspaper Kiosk. Green and black with poster adverts which could vary. Price 1/9d. Issued 1955. Deleted 1970. 4.1cms.	£4	£3	£2

MODEL	MB	MU	GC
R70. Level Crossing. Grey, white, black and rose pink complete with lamps. Price 6/6d. Issued 1955. Deleted 1970. 17.1cms.	£9	£7	£5
R71. Footbridge. Grey, white and black with lemon doors and posters which could vary. One of the rare famous adverts is 'Pedigree Dolls' in white, blue, black and gold. Price 8/11d. Issued 1955. Deleted 1970. 21.9cms.	£9	£7	£5
R72. Gate Keeper's Hut. Grey, yellow and black. Price 1/3d. Issued 1955. Deleted 1970. 6.7cms.	£3	£2	£1
R73. Island Platform Canopy. Golden yellow and white or orange and white. Price 1/6d. Issued 1955. Deleted 1970. 13.3cms.	£3	£2	£1
R74. High Level Pier. Grey and white. Price 9d. Issued 1955. Deleted 1970. Height 7.6cms. approx.	£1	75p	50p
R75. Water Tower. Red, black and grey with yellow doors and windows. Price 2/6d. Issued 1959. Deleted 1970. 8.9cms.	£4	£3	£2
R76. Engine Shed. red, grey and white with yellow or orange windows. Fitted with special lugs so that several units can be grouped together. Price 4/11d. Issued 1955. Deleted 1970. 15.9cms.	£6	£5	£4
R77. Bridge Support. Grey and white. Price 3/11d. Issued 1955. Deleted 1970. Height 7.6cms.	£3	£2	£1
R78. Girder Bridge (large). Grey and orange. Price 25/-. Issued 1956. Deleted 1970. 38.1cms.	£9	£7	£5
R78A. Girder Bridge (small). Grey and orange. Price 12/-. Issued 1956. Deleted 1970. 24.1cms.	£5	£4	£3
R78C. Girder Bridge (with supports). Grey and orange. Price 15/6d. Issued 1968. Deleted 1970. 38.1cms.	£10	£8	£6
R79. Inclined Piers. Grey and white. Made to form a ramp to lead up to bridge R78 or operating hopper car set R161. Price 2/6d. Issued 1956. Deleted 1970. 7.6cms.	£5	£4	£3
R80. Station Set. Contains parts R60, R63, R64L, R64R and R67. Price 21/-. Issued 1955. Deleted 1970.	£15	£12	£10
R81. Main Line Station Set. Contains two R63s, one R64L and one R64R, one R60, one R67 and two R68s. Price 32/6d. Issued 1956. Deleted 1970.	£20	£15	£10
R82. Hopper Unloading Bridge. Grey and white. Price 4/6d. Issued 1956. Deleted 1970. 25cms.	£4	£3	£2

MODEL	MB	MU	GC
R83. Track Bumper. Black. Made for T.C. Series Rolling Stock. Price 1/6d. Issued 1957. Deleted 1970. 6.9cms.	£1	75p	50p
R84. Three Line Side Buildings. Grey, blue, white and black with yellow doors. Price 2/11d. Issued 1956. Deleted 1970. Approximate sizes, 3cms, 5cms and 7cms.	£6	£5	£4
R85. Two Arched Piers. Grey. Price 1/6d. Issued 1960. Deleted 1970. Height 7.9cms.	£1	75p	50p
R86. Telegraph Pole. Grey and black. Price 1/11d. Issued 1956. Deleted 1970. Height 9.5cms. approx.	£1	75p	50p
R87. Loading Gauge. Black and grey. Price 1/11d. Issued 1956. Deleted 1970. Height 7.9cms.	£1	75p	50p
R88. Water Crane. Black, grey and orange. Price 1/11d. Issued 1956. Deleted 1970. Height 8.3cms.	£1	75p	50p
R89. Elevated Track Side Walls. Grey and black. These elevated side walls were made for clipping onto the sides of straight tracks used on incline R79. Price 1/11d. Issued 1956. Deleted 1970. 18.4cms.	£2	£1	50p
R90. Set of Curved Track Side Walls. Grey and black. Price 3/-. Issued 1956. Deleted 1970. 34.3cms.	£2	£1	50p
R93. Curved Track. Grey and black. Price 2/3d. Issued 1955. Deleted 1970. Radius 34.3cms.	£2	£1	50p
R96.Straight Track. Grey and black. Price 1/-. Issued 1955. Deleted 1970. 18.4cms.	£1	75p	50p
R97. Power Connecting and Uncoupling Track. Grey and black. Price 1/6d. Issued 1955. Deleted 1970. 18.4cms.	£2	£1.50	£1
R100. Diamond Crossing. Black and grey. Price 3/11d. Issued 1955. Deleted 1970. 18.4cms.	£3	£2	£1
R101. Left Hand Point. Black and grey. Price 4/6d. Issued 1955. Deleted 1970. 18.4cms.	£3	£2	£1
R102. Right Hand Point. Grey and black. A hand operated point in the same manner as R101. Price 4/6d. Issued 1955. Deleted 1970. 18.4cms.	£3	£2	£1
R104. Half Curved Track. Black and grey. Price 1/11d. Issued 1955. Deleted 1970. Radius 34.3cms.	£1	75p	50p
R105. Quarter Straight Track. Black and grey. Price 6d. Issued 1955. Deleted 1970. 4.8cms.	50p	30p	20p

MODEL	MB	MU	GC
R106. Curved Track.			
Black and grey. Price 1/-. Issued 1955. Deleted 1970. Radius 43.5cms.	£1	75p	50p
R107. Straight Track with Television Interference Suppressor.			
Grey and black. Price 1/11d. Issued 1956. Deleted 1970. 18.4cms.	£2	£1	50p
R108. Eight Straight Track.			
Grey and black. Price 3d. Issued 1955. Deleted 1970. 2.4cms.	50p	30p	20p
R109. Half-Curved Track.			
Black and grey. Price 1/-. Issued 1955. Deleted 1970. Radius 43.5cms.	75p	50p	30p
R140. Signal Gantry.			
Black, white, grey, red and green. A hand-operated gantry. Price 10/6d. Issued 1960. Deleted 1970. 18.4cms.	£10	£7	£5
R142/D. Distant Junction Signal.			
Black, white, grey, yellow, green and red. Price 4/6d. Issued 1960. Deleted 1970. Height 14.6cms.	£4	£3	£2
R142/H. Home Junction Signal.			
Black, white, grey, yellow green and red. Price 4/6d. Issued 1960. Deleted 1970. Height 14.6cms.	£4	£3	£2
R143. Loading Gauge.			
Black and white. Price 1/11d. Issued 1960. Deleted 1970. Height 7.9cms.	£1	75p	50p
R145. Modern Signal Box.			
Brick red, fawn and grey with 'Ipswich' sign. Price 6/6d. Issued 1961. Deleted 1970. 11.9cms.	£5	£4	£3
R146. Modern Engine Shed.			
A double track building in red, fawn, grey and cream. Price 12/6d. Issued 1960. Deleted 1970. 18.74ms.	£9	£7	£5
R147. Refreshment Kiosk.			
Slatey grey, with 'Refreshments' sign in white. Price 2/11d. Issued 1960. Deleted 1970. 8.2cms.	£3	£2	£1
R148. Platform Accessories.			
Green and black. Consists of a Planet tractor, a trolley, a barrow and a weighing machine. Price 5/11d. Issued 1960. Deleted 1970.	£5	£4	£3
R149. Cattle Crossing and Gates.			
Grey and black. Consists of a track crossing, a five bar gate, a double cattle gate, two gate bases and two notice boards with the wording 'Beware of the Trains' in white lettering. Price 3/-. Issued 1960. Deleted 1970.	£3	£2	£1
R170. Level Crossing.			
Grey, white and red, this level crossing is electically operated by remote control as R171. Built in track. Price 10/6d. Issued 1956. Deleted 1970. 18.4cms.	£10	£8	£6

MODEL	MB	MU	GC

R171. Transcontinental Level Crossing (electric).

Grey, white and black, this level crossing is electrically operated by remote control 15 volts AC. It is operated by RT48 red lever frame section. The RT44 black lever frame section is not suitable for this model. Price 11/6d. Issued 1956. Deleted 1970. 18.4cms.

	£10	£8	£6

R172. Set of Five Mile Posts.

White with black numbers. Price 1/6d. Issued 1960. Deleted 1970.

	£1	75p	50p

R173. Set of Four Gradient Posts

White with black lettering. Price 1/6d. Issued 1960. Deleted 1970.

	£1	75p	50p

R175. Level Crossing (single track).

White, grey and red. A hand-operated level crossing. Price 5/6d. Issued 1960. Deleted 1970. 13.3cms.

	£5	£4	£3

R176. Tunnel Single Track.

Brown, green, yellow and black. Price 7/6d. Issued 1959. Deleted 1970. 41.9cms.

	£5	£4	£3

R179. Curved Track Side Walls.

Grey and black, large radius. Price 1/3d. Issued 1960. Deleted 1970. Made for 43.5cms. radius track.

	75p	50p	25p

R180. Viaduct.

Rust red or rose pink. Price 7/11d. Issued 1959. Deleted 1970. 33.3cms.

	£5	£4	£3

R181. High Level Embankment

Brown, blue, grey and black scenic livery. Price 3/11d. Issued 1960. Deleted 1970. 35.6cms.

	£5	£4	£3

R182. Set of Two Cuttings.

Green, brown, black and fawn scenic livery. Price 4/6d. Issued 1960. Deleted 1970. 40.2cms.

	£5	£4	£3

R183. Inclined Embankment First Rise.

Black, brown, yellow, green and fawn scenic livery. Price 3/6d. Issued 1960. Deleted 1970. 42.8cms.

	£2	£1	50p

R184. Inclined Embankment Second Rise.

Fawn, black, grey, brown and blue scenic livery. Price 3/11d. Issued 1960. Deleted 1970. 42.8cms.

	£2	£1	50p

R185. Inclined Embankment Third Rise.

Fawn, black, grey, brown and blue scenic livery. Price 4/-. Issued 1960. Deleted 1970. 42.8cms.

	£3	£2	£1

R186. Field Fencing.

White. Price 2/-. Issued 1960. Deleted 1970.

	£1	75p	50p

R187. Pedestrian Crossing.

Black and grey livery. Consists of track crossing, double swing gates, fence, single swing gate, fence, two gate bases and two notice boards with the words 'Beware of the Trains' in white. Price 2/11d. Issued 1960. Deleted 1970.

	£2	£1	50p

R188. River Bridge.

Lime green and white. Price 3/6d. Issued 1951. Deleted 1970. 33.3cms.

	£5	£4	£3

MODEL	MB	MU	GC

R189. Brick Bridge.
Red. Price 2/11d. Issued 1961. Deleted 1970. 11.1cms.

	£2	£1.50	£1

R201. Left Hand Point (electric).
Black and grey. Price 3/6d. Issued 1956. Deleted 1970. 18.4cms.

	£4	£3	£2

R202. Right Hand Point (electric).
Black and grey. Price 3/6d. Issued 1956. Deleted 1970. 18.4cms.

	£4	£3	£2

R263/L Point Operated Signal.
Black, white, red and green. This is a left hand point operated
signal. Price 4/11d. Issued 1955. Deleted 1970. Height 10.8cms.

	£3	£2	£1

R263/R. Point Operated Signal.
Black, white, red and green. This is a right hand point operated
signal. Price 4/11d. Issued 1965. Deleted 1970. Height 10.8cms.

	£3	£2	£1

R264. Grand Victorian Suspension Bridge.
Lime green, dark blue, grey, orange and white. consists of two end
piers, two high central piers, four lengths of bridge span and eight
suspension units. Price 39/6d. Issued 1963. Deleted 1970. Full
length 138.5cms.

	£25	£20	£15

R266. Pair of Station Lamps.
Black, white and red. Price 13/9d. Issued 1965. Deleted 1970.

	£5	£4	£3

R167. Fog Signal.
Grey-blue and black. As train passes hut a 'maroon' is automatically
fired. Mechanically operated and special caps are used. Price 4/6d.
Issued 1965. Deleted 1970.

	£3	£2	£1

R268. Bell Signal.
Grey, white and black. Bell signal operates from 12/15 volts
AC/DC which enables a train to be passed from one track zone to
another using authentic railway bell signal procedure. Price 6/11d.
Issued 1965. Deleted 1970.

	£5	£4	£3

R269. Suspension Bridge Extension Set.
Lime green containing two end piers, one high central pier, two
lengths of bridge span and four suspension units. Also in blue grey
and orange. Price 24/-. Issued 1963. Deleted 1970.

	£6	£5	£3

R360. Windmill.
Yellowy green with blue door. The first of a series of buildings.
Made specifically to decorate any railway layout. Price 4/6d. Issued
1960. Deleted 1970. Height 20.9cms.

	£6	£5	£3

MODEL	MB	MU	GC
R361. Church.			
Dark brown, green, fawn and black. Price 4/6d. Issued 1960. Deleted 1970. 13.9cms.	£5	£4	£3
R362. Oast House.			
Fawn, a pinkish white, blue and black. Price 3/11d. Issued 1960. Deleted 1964. Height 15.2cms.	£3	£2	£1
R363. Forge.			
Pinkish white, fawn and black. Price 2/11d. Issued 1960. Deleted 1964. 12.7cms.	£3	£2	£1
R364. Gas Holder.			
Black brick design. Price 2/3d. Issued 1960. Deleted 1964. Diameter 10.8cms.	£3	£2	£1
R365. Barn.			
Dark brown, black, pink and grey. Price 2/6d. Issued 1960. Deleted 1964. Length 12.7cms.	£2	£1.50	£1
R366. Pair of Haystacks.			
Fawn. Price 1/11d. Issued 1960. Deleted 1964. Height 7cms.	£1	75p	50p

MODEL	MB	MU	GC
R367. Thatched Cottage.			
Fawn, pink, blue and black. Price 4/6d. Issued 1960. Deleted 1964. Length 10.8cms.	£4	£3	£2
R368. Inn.			
Pinkish white, brown and black. Price 4/6d. Issued 1960. Deleted 1964. Length 12.7cms.	£5	£4	£3
R369. Village Store.			
Pinkish white, brown and black. Price 4/11d. Issued 1960. Deleted 1964. Length 12.7cms.	£6	£5	£4
R370. Farm Cottage.			
Fawn, dark brown, green and black with a blue door. Price 4/6d. Issued 1961. Deleted 1964. Length 9.5cms.	£7	£5	£3
R371. Factory.			
Dark fawn, grey and black with green door. Price 4/6d. Issued 1961. Deleted 1964. 12.7cms.	£5	£4	£3

MODEL	MB	MU	GC

R372. Oil Tank.
Silver grey tank with 'Shell' decals in large gold or orange letters.
Price 3/6d. Issued 1961. Deleted 1964. Diameter 9.5cms.

£4 £3 £2

R373. Coal Dump.
Off white, dark brown with imitation black coal and 'James & Son
Ltd. Coal Merchant' in white lettering on black background. Price
2/11d. Issued 1961. Deleted 1964. Length 10.2cms.

£3 £2 £1

R374. Vicarage.
Dark brown, fawn, blue and grey. Price 4/11d. Issued 1961.
Deleted 1964. Length 11.4cms.

£5 £4 £3

R375. Terraced House.
Pink, dark brown and black. Price 4/6d. Issued 1961. Deleted 1964.
Length 11.1cms.

£4 £3 £2

R376. Countryside Accessory Pack.
Grey, black and blue, consisting of a TV aerial, a weather cock, a
lightning conductor, an 'Inn' sign, store name board and ladder.
Price 4/11d. Issued 1961. Deleted 1964.

£3 £2 £1

R377. Large Fir Tree.
Dark brown and green. Price 3/6d. Issued 1961. Deleted 1964.
Height 13.6cms.

£3 £2 £1

R378. Pair of Small Fir Trees.
Dark brown and green. Price 4/-. Issued 1961. Deleted 1964.
Height 9.8cms.

£3 £2 £1

R379. Large Conker Tree.
Brown and green. Price 3/-. Issued 1961. Deleted 1964. Height
13.6cms.

£3 £2 £1

R380. Pair of Small Conker Trees.
Brown and green. Price 3/2d. Issued 1961. Deleted 1964. Height
9.8cms.

£3 £2 £1

R394. Hydraulic Buffer.
Yellow and black, spring loaded. Price 3/11d. Issued 1961. Deleted
1970. 3.8cms.

£3 £2 £1

R412. Quadruple Push Button Control.
Grey and orange. Made to control four electrically operated points,
complete with leads. Price 5/11d. Issued 1964. Deleted 1970.

£3 £2 £1

R414. Double Curve Level Crossing.
This hand-operated crossing fits double curve track sections of
Super 4. Grey white and brown. Price 8/6d. Issued 1964. Deleted
1970. 18.4cms.

£15 £12 £9

R430. Small Radius Sidewalls.
Orange or dark brown, these curved sidewalls fit to R483 and
R484. Price 2/6d. Issued 1963. Deleted 1970.

£1 60p 40p

R431. Large Radius Curved Sidewalls.
Orange or dark brown. Fit R485, and R486 tracks. Price 2/9d.
Issued 1963. Deleted 1970.

£1 75p 50p

MODEL	MB	MU	GC
R432. Girder Bridge Set.			
Contains R77, bridge supports, R78 girder bridge, 2 x R453 high level piers, 2 x R457 inclined piers, 2 x R456 straight sidewalls and R173 gradient posts. Price 39/6d. Issued 1959. Deleted 1970. 22.7cms.	£7	£6	£5
R433. Underlay for R490.			
Price 1/3d. Issued 1963. Deleted 1970.	75p	50p	25p
R434. Underlay for R491.			
Price 1/3d. Issued 1963. Deleted 1970.	75p	50p	25p
R435. Underlay for R492.			
Price 1/3d. Issued 1963. Deleted 1970.	75p	50p	25p
R436. Underlay for R493.			
Price 1/3d. Issued 1963. Deleted 1970.	75p	50p	25p
R437. Hand Operated Y Point.			
Brown and grey. Price 7/6d. Issued 1964. Deleted 1970. Radius 43.8cms.	£5	£4	£3
R438. Underlay for R437.			
Price 1/3d. Issued 1963. Deleted 1970.	75p	50p	25p
R453. High Level Pier.			
Grey or dark orange. Price 1/6d. Issued 1963. Deleted 1970. Hieght 7.3cms.	50p	30p	20p
R454. Siding Set.			
Grey or dark brown. Original price 14/-. Issued 1964. Deleted 1970.	£3	£2	£1
R455. Mast Base.			
Brown, black, white and blue. Price 3½d. Issued 1964. Deleted 1970.	£1	50p	30p
R456. Straight Sidewalls.			
Grey or orange. To fit to R480, R481 and R489. Price 2/6d. Issued 1963. Deleted 1970.	60p	40p	20p
R457. Set of 7 Inclined Piers.			
Dark brown or orange. Price 4/11d. Issued 1963. Deleted 1970.	£3	£2	£1
R460. Straight Platform Unit.			
White and grey. Price 2/9d. Issued 1960. Deleted 1970. 16.8cms.	£2	£1	50p
R461. Straight Platform Unit with Subway.			
Grey and white with K.L.G. advert on yellow backgound. Price 3/6d. Issued 1963. Deleted 1970. 16.8cms.	£3	£2	£1
R462. Large Radius Curved Platform Unit.			
Grey and white. Fits round the outer edge of Super 4 and Series 3 large radius curved tracks. Price 2/11d. Issued 1962. Deleted 1970. 16.8cms.	£3	£2	£1
R463. Small Radius Curved Platform Unit.			
Grey and white. Price 2/6d. Issued 1962. Deleted 1970. 7.6cms.	£1	75p	50p

MODEL	MB	MU	GC

R464. Platform Ramp.
Double curved ramp in white and grey. Price 1/9d. Issued 1962.
Deleted 1970. 12.8cms.

	£1	75p	50p

R465. Steps Unit.
Grey and white. Price 3/6d. Issued 1962. Deleted 1970. 16.8cms.

	£3	£2	£1

R466. Straight Canopy Set.
Golden yellow or orange. Price 2/11d. Issued 1962. Deleted 1970.
16.8cms.

	£2	£1.50	£1

R467. Large Radius Curved Canopy Set.
Golden yellow or orange. Price 3/6d. Issued 1962. Deleted 1970.
16.8cms. approx.

	£2	£1.50	£1

R468. Small Radius Curved Canopy Set.
Golden yellow or orange. Fits to R463 small radius curved platform
unit. Price 2/9d. Issued 1962. Deleted 1970. 7.8cms. approx.

	£2	£1.50	£1

R470. Name Boards.
A selection of adhesive names in black and white. Price 1/-. Issued
1961. Deleted 1970.

	50p	30p	20p

R471. Platform Fencing Unit.
Golden yellow or orange. Price 2/6d. Issued 1962. Deleted 1970.
16.8cms.

	£2	£1.50	£1

R472. Island Platform Waiting Room.
Grey, black and yellow. Price 3/11d. Issued 1963. Deleted 1970.
15.8cms. approx.

	£3	£2	£1

R473. Ticket Office.
Red, black, yellow and white. Price 12/6d. Issued 1961. Deleted
1970. 16.8cms.

	£7	£6	£5

R474. Station Upper Floor With Clock Tower.
Dark brown, black, yellow and white. This upper floor fits above
two ticket offices mounted side by side. Price 14/6d. Issued 1961.
Deleted 1970. 24.1cms.

	£15	£12	£10

R477. Place Names.
Three sets of four names, self-adhesive labels. Place names. Price
1/-. Issued 1961. Deleted 1970.

	£1	75p	50p

MODEL	MB	MU	GC

R478. Set of Six Telegraph Poles.
Black with clip fit for Super 4 and Series 3 track. Price 3/11d.
Issued 1964. Deleted 1970. Height 9.5cms.

	£2	£1.50	£1

R479. Loading Gauge.
Black and white. Made to clip fit to Super 4 and Series 3 track.
Price 1/3d. Issued 1964. Deleted 1970. Height 7.6cms.

	£1	75p	50p

R480. Double Straight Track.
Brown and grey. Price 1/9d. Issued 1964. Deleted 1970. 33.5cms.

	£1.50	£1	50p

R481. Straight Track.
Brown and grey. Price 1/-. Issued 1964. Deleted 1970. 16.8cms.

	£1	75p	50p

R482. Quarter Straight.
Brown and grey. Price 9d. Issued 1964. Deleted 1970. Length
3.8cms.

	50p	30p	20p

R483. Double Curved Track.
Brown and grey. Price 1/9d. Issued 1964. Deleted 1970. Radius
37.2cms.

	£1.50	£1	50p

R486. Curved Track.
Brown and grey. Price 1/-. Issued 1964. Deleted 1970. Radius
37.3cms.

	£1	75p	50p

R485. Double Curved Track.
Brown and grey. Price 1/9d. Issued 1964. Deleted 1970. Radius
43.8cms.

	£2	£1.50	£1

R486. Curved Track.
Brown and grey. Price 1/-. Issued 1964. Deleted 1970. Radius
43.8cms.

	£1	75p	50p

R487. Power Connecting Clip.
Brown, complete with television interference suppressor. Price
1/9d. Issued 1964. Deleted 1970.

	£1	75p	50p

R488. Uncoupling Ramp.
Brown, black and orange. Price 1/-. Issued 1964. Deleted 1970.

	50p	40p	20p

R489. Straight Long Track.
Brown and grey. Price 3/-. Issued 1964. Deleted 1970. 67cms.

	£3	£2	£1

R490. Left Hand Point.
Brown and grey. Hand-operated. Price 7/6d. Issued 1964. Deleted
1970. Radius 43.8cms.

	£6	£5	£4

R491. Right Hand Point.
Brown and grey. Hand-operated. Price 7/6d. Issued 1964. Deleted
1970. Radius 43.8cms.

	£6	£5	£4

R492. Left Hand Diamond Crossing.
Brown and grey. Price 7/6d. Issued 1964. Deleted 1970. 16.8cms.

	£3	£2	£1

R493. Right Hand Diamond Crossing.
In brown and grey. Price 7/6d. Issued 1964. Deleted 1970.
16.8cms.

	£3	£2	£1

MODEL	MB	MU	GC
R494. Buffer Stop. Orange and brown. Made for Super 4 and Series 3 track. Price 1/-. Issued 1964. Deleted 1970.	£2	£1.50	£1
R495. Single Track Level Crossing. A single hand-operated crossing in grey, white and brown. Price 6/6d. Issued 1964. Deleted 1970. 9.8cms.	£7	£5	£3
R496. Double Track Level Crossing. Double hand-operated crossing in grey, white and brown. Price 7/11d. Issued 1964. Deleted 1970. 14.5cms.	£10	£8	£6
R497. Isolating Track. Brown and grey. Price 2/6d. Issued 1964, Deleted 1970. 16.8cms.	£3	£2	£1
R520. Smoke Unit. The most realistic device ever to be fitted to British built 00/HO gauge locomotives. Place a few drops of smoke oil from the capsule provided down the chimney of the locomotive fitted with the smoke unit and once the locomotive is in motion and the unit has warmed up, smoke is emitted from the chimney. A standard fitting on all or most of the models which the Triang Hornby Company built after the 1st January 1961. A letter S on the underside of the chassis denotes that locos are suitable for this fitting. Price 17/6d. Issued 1961. Deleted 1970.	£9	£7	£5
R521. Capsule of Smoke Oil. Price 1/3d. Issued 1961. Deleted 1970.	50p	30p	20p
R573. Colour Light Signal Gantry Set. Black, white and yellow, set operates from 12/15 volts AC/DC bases clip fit to track. It includes two lever frame sections which can be operated by automatic train control sets. Price 29/6d. Issued 1966. Deleted 1970.	£9	£7	£5
R576. Tunnel. Single track tunnel is green, brown, white and cream. Price 10/6d. Issued 1964. Deleted 1970. 25.4cms.	£5	£4	£3
R580. Pair of Gantries. Lime green and brown. Made to clip fit to double tracks to support pairs of catenary wires above the rails. Price 4/11d. Issued 1966. Deleted 1970.	£3	£2	£1
R581. Signal Accessory Pack. Price 16/11d. Issued 1961. Deleted 1970.	£5	£4	£3
R582. Ticket Office and Platform Unit. Grey, brown, black, yellow and green. Unit consists of two R148 Platform Accessories, R460 Platform unit, R465 steps unit and R473 Ticket Office. Price 14/6d. Issued 1963. Deleted 1970.	£9	£7	£5
R583. Large Radius Curved Platform with Canopy. Grey, blue, white, green and cream. Price 5/11d. Radius 16.8cms approx.	£5	£4	£3
R584. Small Radius Curved Platform with Canopy. Grey, blue, white, green and cream. Price 4/11d. Issued 1963. Deleted 1970. Radius 9.1cms approx.	£3	£2	£1

MODEL	MB	MU	GC
R585. Straight Platform Canopy and Seat Unit. Grey, blue, white, green and cream. Price 5/11d. Issued 1963. Deleted 1970. 15cms.	£6	£4	£2
R586. Two Platform Fencing Units. Yellow and white or cream and yellow. Consisting of platform fencing unit and name boards with a selection of station names. Price 2/6d. Issued 1963. Deleted 1970.	£2	£1.50	£1
R587. Trackside Accessory Pack. Grey, or white and black. Consisting of five mile posts, four gradient posts and two whistle signs. Price 2/11d. Issued 1963. Deleted 1970.	£2	£1.50	£1
R588. Island Platform Set. Yellow, maroon, white, grey and blue. Price 16/-. Issued 1965, Deleted 1970. Length 59cms.	£6	£5	£4
R5005. Engine Shed Kit. Grey, mustard and green. Kit was easily assembled from individual parts and could be extended to cover four tracks. Price 10/6d. Issued 1968. Deleted 1970.	£6	£5	£4
R5006. Engine Shed Extension Kit. Grey, mustard and green. Price 5/-. Issued 1968. Deleted 1970.	£3	£2	£1
R5015. Girder Bridge. Orange and grey. Price 4/11d. Issued 1966. Deleted 1970. 16cms approx.	£5	£4	£3
R5020. Goods Depot Kit. Grey, fawn and green, with working crane in orange, black and fawn. Price 9/6d. Issued 1968. Deleted 1970. 16cms approx.	£10	£7	£5
R5030. Island Platform Kit. Fawn, grey and black. Price 15/6d. Issued 1968. Deleted 1970. 34cms approx.	£10	£7	£5
R5083. Terminus and Through Station Kit. Fawn, black and white with green windows. Price 15/6d. Issued 1968. Deleted 1970.	£15	£12	10
R5084. Canopy Extension Kit. Fawn, black and white, Parts screw together simply and quickly. Price 10/6d. Issued 1968. Deleted 1970.	£5	£4	£3
R5086. Platform Extension. Grey and black. Price 2/11d. Issued 1967. Deleted 1970. 9.8cms.	£1.50	£1	50p
R5087. Platform Fence. Fawn. This fits 5086 and 5089. Price 1/6d. Issued 1967. Deleted 1970. 9.8cms.	50p	30p	20p
R5089. Side Platform Extension. Fawn. Price 1/-. Issued 1967. Deleted 1970. 9.8cms.	50p	30p	20p
R5092. Double Track Tunnel. Green and fawn. Price 7/6d. Issued 1967. Deleted 1970. 15.8cms approx.	£6	£5	£4

TRI-ANG TRAIN SETS

MODEL	MB	MU	GC

RP.A. Train Set Clockwork.
Contains Clockwork R.255 Saddle Tank loco in black and two
R.230 coaches in maroon, black, and grey, lapel badge and curved
and straight track to make a radius 71cms x 91cms. Price 25/-.
Issued 1960. Deleted 1969. £40

With black and green loco, otherwise as above. £25

RP.B. Clockwork Train Set.
Contains R.256 diesel shunter in maroon and grey, two R.217 open
trucks in pea green or light lime green, R.218 closed van in green
and grey, lapel badge and a small radius track to make an oval
71.1cms. Price 21/-. Issued 1961. Deleted 1969. £25

RP.C. Train Set.
An electric set containing R.252 steeple cab loco in maroon and
grey with non-operating pantograph, two R.231 coaches in green,
grey and white, a power-connecting clip, lapel badge and a small
radius curved track to make an oval 71.1cms Price 29/11d. Issued
1961. Deleted 1968. £40

Also with green loco. Very rare. Otherwise as above. £75

RP.E. Electric Goods Set.
Contains R.359 Tank Loco in black or grey, latter worth double,
two R.217 Open trucks in green, R.218 closed van in green, grey
and white, power-connecting clip, Lapel badge, plus a small radius
circle track to form oval 71.1cms. Price 29/6d. Issued 1961. Deleted
1968. £30

Also with green loco. Rare. Otherwise as above. Primary Series
Loco sets were worked from 12/15 volts D.C. for battery operation.
A battery connecting unit and three 4½ volt dry batteries were
required for operation. From AC mains electric one needed a power
controller having an output of 12/15 volts D.C. £50

R.P. Clockwork Set.
One of the first sets produced by the Triang Company. A fine
passenger set containing a black R151 loco, one R120 coach, one
R121 coach, both in maroon and cream with grey roof, and a
circular track. Price 27/6d. Issued 1955. Deleted 1960. £15

R.Q. Clockwork Set.
A high quality goods set containing a black loco 154, one R10
truck in grey, one R11 van in dark brown and cream, one R19
wagon in green and black and one R16 guards van in maroon,
cream and grey, plus a circular track. Price 35/-. Issued 1955.
Deleted 1960. £15

R.A. Electric Set.
A fine passenger set containing one black R50 loco and matching
R30 tender. One R28 coach and one R29 coach in maroon, cream,
grey and black, complete with RT 41 battery box and oval track.
Original price 45/-. Issued 1955. Deleted 1965. £20

R.A.X. Electric Set.
This set contains exactly the same as the previous R.A. set, the
only difference being it contains the R.T. 42 speed control unit in
place of the R.T. 41 battery box. Price 50/-. Issued 1956. Deleted
1966. £25

R.B. Electric Set.

Contains green and black R53 loco and matching R31 tender. One R28 coach and one R29 coach in matching maroon, cream and grey livery. Price 50/-. Complete with R.T. 41 battery box and oval track. Issued 1955. Deleted 1965.

£20

R.B.X. Electric Set.

High quality set contains the same as the R.B. set but with the R.T. 42 speed control unit in place of the R.T. 41 battery box. Price 55/-. Issued 1956. Deleted 1966.

£25

R.C. Electric Set.

Contains black R59 loco, one R120 and one R121 passenger coaches in matching maroon, yellow and grey and black complete with R.T. 41 battery box and oval track. Price 57/6d. Issued 1955. Deleted in 1965.

£25

R.C.X. Passenger Electric Set.

Set contains the same as R.C. but with an R.T. 42 speed control unit in place of the R. T. 41 battery box. Price 59/-. Issued 1956. Deleted 1966.

£27.50

R.D. Goods Electric Set.

Contains one black R59 loco, one orange and cream R118 wagon, one R12 wagon in silver with B.P. decals in blue, green and red, one R110 wagon in grey and black and one R16 van in dark brown, yellow and red. Price 61/-. Issued 1955. Deleted 1965.

£20

R.D.X. Goods Electric Set.

Set contains the same as R.D. but with R.T. 42 speed control unit in place of the R.T. 41 battery box. Price 65/-. Issued 1956. Deleted 1966.

£25

R.E. Goods Electric Set.

Contains one black R52 loco with red motif, one grey, white and black R18 wagon, one R12 Shell wagon in silver with red and yellow decals, one R10 truck in brown and black with N.E.' decals in yellow and one R16 van in black and dark brown. Complete with one R.T. 41 battery box and oval track. Price 60/-. Issued 1955. Deleted 1965.

£20

R.E.X. Goods Electric Set.

Set contains the same as that of the previous R.E. set but with R.T. 42 speed control unit in place of the R.T. 41 battery box. Price 63/-. Issued 1956. Deleted 1966.

£25

R.H.X. Continental Diesel Set.

Silver, orange and grey with yellow tint. This set consists of one R55 diesel loco, and one R57 diesel dummy end, one R24 coach, one R25 coach, an RT 42 speed control unit and oval track. Price 63/-. Issued 1956. Deleted 1966.

£25

R.J.X. Box Car Set.

Contains black and grey R54 loco with matching R32 tender, one R116 gondola wagon in green, one R114 box car in red, black and yellow and one R115 caboose in matching livery. One R.T. 42 speed control unit and oval track. Price 60/-. Issued 1956. Deleted 1966.

£25

R.K.X. Hopper Goods Set.

Contains one R56 loco in black, one R111 hopper car in green, one R117 oil tanker in orange and black, one R116 gondola wagon in grey, one R115 caboose in red, grey and black, one R.T. 42 speed control unit and a set of oval track. Price 60/-. Issued 1956. Deleted in 1966.   £30

R.L.X. Diesel Goods Set.

Contains one R55 diesel loco in orange, yellow and silver grey with one R57 diesel dummy end in matching livery. One R116 gondola in blue and orange with white trim, one R114 box car in green, black and grey, one R115 caboose in red, grey and black, one R.T. 42 speed control unit and oval track. Price 65/-. Issued 1956. Deleted 1966.   £25

R.45. Turntable Set.

Electrically operated from 12 volt DC by remote control. Brown, white, green, yellow and red with black and yellow leads. Also in silver grey. Diameter 33.7 cms. Price 17/6d. Issued 1956. Deleted 1966.   £7

R80 Station Set.

Contains a station building in maroon, cream, grey and black, consisting of parts R60, R63, R64L, R64R and R67. Price 12/6d. Issued 1955. Deleted 1965.   £7

R81 Main Line Station Set.

Contains station buildings, seats and advertising boards in maroon, grey, lemon, blue and black on grey and white platform. This set consists of two R63, one R64L, one R64R, one R60, one R62, one R67, and two R68. Price 21/-. Issued 1956. Deleted 1966.   £10

R.161 Operating Hopper Car Set.

Contains one R111 hopper car in green and yellow with Triang decals, one R82 hopper unloading bridge in grey, two R74 high level piers in grey, one R85 set of arched piers in grey, one R162 brush and shovel in black and one R163 Mineral bunker, complete with imitation minerals. Packed in a fine display box, it made a very interesting gift. A locomotive draws the loaded hopper car up an incline constructed with this set and when the hopper car reaches the unloading bridge a trip releases the door plate in the base of the hopper, allowing the mineral to fall through, either into another truck awaiting below, or back into the mineral bunker. Price 21/-. Issued 1956. Deleted 1966.   £15

R.F. Train Set.

Lime green, grey and black, this fine passenger set consists of R156 motor coach and R225 dummy end and one RT41 battery box complete with oval track. Price 35/-. Issued 1957. Deleted 1967.   £25

R.F. Special 4 Car Suburban Set.

Typical Southern Region four-car suburban electric train set made up with two R156 units and two R225 units in authentic green, black and grey livery. Also containing an R. T. 42 speed control unit. In special presentation box. Price 95/-. Issued 1957. Deleted 1967.   £45

R.F.X. Train Set.

Same as the previous R.F. set but with an R.T. 42 speed control unit in place of R.T. 41 battery box. Price 42/6d. Issued 1957. Deleted 1967.   £30

MODEL	MB	MU	G(

R.S.1. Main Line Passenger Train Set.

Contains the Princess Victoria R50 engine in black or grey with
authentic British Rail emblems and decals. With an R30 matching
tender, one R29 composite coach and one R28 brake second coach
in matching maroon, lemon, grey and black. One R197 power
connection clip, one R special presentation box and a set of straight
and curved track, plus uncoupling tracks forming an oval 91 x
71 cm. Price 79/6d. Issued 1958. Deleted 1968.   £35

RS.2. The Princess Royal Set.

Maroon, grey and black with decals and numbers. Consisting of
Loco, R. 258 Princess Royal, R. 34 tender, 2 x R. 321 composite
coaches, R. 320 brake 2nd coach, R. 197 connecting clip. With
straight and curved track, plus uncoupling track, forming an oval
109 cms x 91 cms. Price £5-7-6. Issued 1958. Deleted 1969.   £30

R.S.3. The Britannia Pullman Set.

Green and black R.259 engine with R. 35 matching tender with
Britannia decals and numbers. Also includes, 2 R. 228 Pullman
first-class cars 1.R. 328 Pullman Brake 2nd car, R. power 197
connecting clip. With straight and curved track, plus uncoupling
track, forming an oval 109 x 91 cms. Price £5-19-6d. Issued 1959.
Deleted 1969.   £40

RS.4.0-6-0 Class 3-F Tank Loco Set.

Contains black 0-6-0 R. 52 Tank loco, R. 10 black or grey open
wagon, R. 14 cream, black and grey fish wagon; R. 16 brown, black
and cream or white brake van; R. 197 power connecting clip. With
straight and curved track to form an Oval 91 x 71 cms. Price
£2-12-6d. Issued 1958. Deleted 1969.   £25

RS.5. Goods Train Set.

Contains a black 0-6-0 loco and matching R. 33 tender, R. 118
Bogie well wagon in grey; R. 29 bogie brick wagon in brown or
chocolate with brick decals; R. 122 cattle wagon in brown, grey and
black; R. 124 W.R. brake van in brown and grey and black; R. 197
power connecting clip, and straight, curved and uncoupling track to
form an oval 91 x 71 cms. Price £2-19-6d. Issued 1959. Deleted
1969.   £25

RS.6. Diesel Shunter Set.

Green consisting of R. 152 diesel shunter, R. 18 cable drum in
black and green with 'Pirelli' decals, R. 13 open wagon with coal
load in black and grey, R. 12 petrol tanker in grey, yellow and red
with 'Shell' decals; R. 16 brake van in brown, grey and black,
R. 197 connector clip and straight, curved and uncoupling track,
forming an oval 91 x 71 cm. Price 63/-. Issued 1960. Deleted 1969. £25

R.S.7. Railcar Diesel Set.

Green, grey, black and grey with yellow or gold trim lines.
Contains R. 157 diesel power car, R. 158 diesel trailer car, R. 197
power connecting clip and straight, curved and coupling track
which forms an oval of 91 x 71 cms. Price 50/-. Issued 1960.
Deleted 1969.   £25

RS.8. The Midlander Set.

Maroon with full B.R. Midland' decals and numbers, contains
locomotive and tender with composite and brake second coaches.
Straight, curved and uncoupling track to make an oval 109 x
91 cms. One of the best investments in the whole Triang train set
range. Price 95/-. Issued 1962. Deleted 1969.   £50

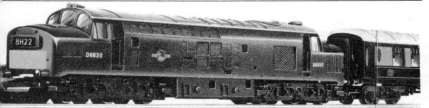

RS.9. Inter City Express.

Contains English Electric Co-Co diesel locomotive in green, black
and yellow with grey roof, two first class Pullman coaches and one
Pullman brake second car in maroon or dark brown, cream, black
with white roofs. Pullman coaches called 'Mary' and 'Anne' and
word 'Pullman' in gold lettering. It contains a large oval track with
uncouples, giving an oval of 109 x 91 cms. Price 129/2d. Issued
1966. Deleted 1973.   £30

RS.10. 'Polly' Goods Set.

Contains red and black 'Polly' loco, a flat wagon in dark brown
containing a green Minic car, a brake van in dark brown, black and
cream, a mineral wagon in grey and black, a 12 volt DC power
controller fitted with speed and directing regulator with a thermal
automatic overload cut out and resistance adjusting switch. Also
section of track to make an oval of 91 x 71 cms. Price 47/6d. Issued
1965. Deleted 1973.   £20

RS.11. Diesel Freight Train Set.

Contains an R353 yard switcher diesel engine in yellow, black and
red, an R237 depressed centre car with low loader containing
bulldozer in green, red, fawn and white, an R117 oil tanker in blue
and black with 'Shell' decals; a 116 gondola in green with white
decals; an R115 caboose in maroon, black and grey, an R197 power
connecting clip complete with straight, curved and uncoupling
track forming an oval 91 cms.

RS.11/A. Diesel Freight Train Set.

Contains R353 yard switcher in green, black and red, R116 gondola
in grey and black; R117 oil tanker in blue with 'Shell' decals; R115
caboose in brown, black and grey; R197 power connecting clip,
complete with straight, curved and uncoupling track, forming an
oval 91 x 71 cms. Price 55/6d. Issued 1961. Deleted 1969.   £25
Price for green, black and red (rare)   £50

RS.12. Freight Set with Pacific Loco.

Contains R54 Pacific loco in dark green and black; three coal
trucks in black and grey and a caboose in maroon, black and grey,
with the rare NCB decals. With a section of straight, curved and
uncoupling track plus power connecting clip to form a large oval
127 x 109 cms. Price 95/-. Issued 1960. Deleted 1969.   £75

RS.12/A. Diesel Freight Set.

Contains R155 diesel switcher in yellow, grey and red; R237
depressed centre car with low loader and bulldozer in green, fawn,
red and white; an R238 depressed centre car in blue containing two
Transatlantic cable drums in green and black with white decals; an
R115 caboose in maroon, black and grey; an R197 power
connecting clip with straight, curved and uncoupling to form an
oval 109 x 71 cms. Price 75/-. Issued 1960. Deleted 1969.   £20

RS.13. Passenger Train Set.

A beautiful set in matching livery of maroon, grey, black and green,
with Triang Railway' decals and numbers. Contains R55 diesel
loco, an R130 baggage car; an R25 Vista dome car; an R24 coach;
an R125 observation car; an R197 power connecting clip complete
with straight, curved and uncoupling track forming an oval 127 x
109 cms. Price £5.7s.6d. Issued 1961. Deleted 1969.   £50

RS.14. Passenger Train Set.

Another beautiful set in matching livery of blue, golden yellow,
black and grey. Contains an R159 double ended diesel loco, an
R134 baggage car, an R131 coach, an R132 vista dome car, an
R133 observation car, an R197 power connecting clip complete
with straight, curved and uncoupling track forming an oval 127 x
109 cms. Price £5.12.6d.   £50

RS.15. Freight Train Smoker Set.

Contains R54 Pacific loco in grey, white or black and white; an
R32 tender in matching grey or black with the number 2335; an
R136 box car in green and black; an R129 refrigerator car in white
and black; an R126 stock car in yellow and black; an R115 caboose
in maroon, black and grey; an R197 power connecting clip
complete with straight, curved and uncoupling track forming an
oval 127 x 109 cms. Price £6.12.6d. Issued 1961. Deleted 1969.   £35

RS.16. Strike Force Set.

Fine military battle space set in matching livery of khaki brown. A
Strike Force Ten set train equipped for ground and air warfare.
Contains synchrosmoking loco; an R562 catapult plane launching
car with catapulting plane in yellow, blue and red which catapults
and flies up to 20 feet. It is equipped with a clip fit launching
trigger; an R568 assault tank transporter with twin missile firing
tank and oval of track with automatic uncoupler to make an oval
127 x 109 cms. Also includes a detachment of twelve battle space
commandos. Price 95/-. Issued 1965. Deleted 1970.   £40

RS.17. Satellite Set.

Set is in red, blue and black livery with yellow, green and white
decals; plus a silver, red and yellow satallite. Train is equipped
with satallite launcher and tracking radar control with this set.
Made in a limited number. Set also includes an R566 Spy Satellite
launching car off which a pre-wound satellite boosts as car passes
the clip fit trackside figure. Also includes an R567 radar tracking
command car with revolving scanner, flashing light and oval of
track with automatic uncoupler to make an oval of 12 x 109 cms
and includes twelve battle space commandos. Price 95/-. Issued
1965. Deleted 1969.   £45

RS.18. Clockwork Action Set.

Set is in matching red, black and grey with a clockwork loco, a
tank carrier with tank and two gun carriers, plus a searchlight
wagon. Rare set, as only a limited number produced. With an oval
track 91 x 71 cms. Price 25/6d. Issued 1964. Deleted 1966.   £75

RS.19. Clockwork Goods Set.

Green, black and gold with milk wagon in white, log wagon in
brown and black, oil tanker in silver and red, and caboose in
maroon and silver; with an oval track 91 x 71 cms. Price 23/1d.
Issued 1965. Deleted 1969.   £40

RS.20. Battery Goods Set.

Red and black with three matching coal trucks and guards van, plus a circular track 91 x 71 cms. Price 57/6d. Issued 1965. Deleted 1969.

£50

RS.21. Main Line Set.

Continental green with gold lines and black lettering. Three carriages, restaurant car and guards van, and large oval track 127 x 109 cms. Price 160/-. Issued 1965. Deleted 1969.

£75

RS.22. Night Sleeper Set.

Beautiful set in matching maroon, black and grey livery, complete with crew and smoke. With large oval track 127 x 109 cms. Price 129/6d. Issued 1963. Deleted 1969.

£45

RS.23. Pullman Set.

Contains loco in lime green, black and yellow with matching Pullman coaches and restaurant car in dark brown, yellow and off white, with crew and smoke. Complete with a large circular track 127 x 109 cms. Price 143/6d. Issued 1963. Deleted 1969.

£40

RS.24. Pick Up Goods Set.

Contains very popular 040 Industrial Locomotive 'Nellie' in blue, black and red; an R14 Fish van in white; an R10 open wagon in green; an R16 brake van in dark brown, black and white; a section of curved, straight and uncoupling track to form an oval layout 99 x 81 cms. Price 57/6d. Issued 1964. Deleted 1969.

£30

RS.24/A. Pick Up Train Set.

Contains the 040 'Polly' loco in red, black and yellow; a locomotive freight van with opening doors in fawn, and grey; an open wagon in grey and black; a brake van in brown, black and grey and an oval track 99 x 81 cms. Price 57/6d. Issued 1968. Deleted 1973.

£35

RS.25. B.R. Goods Train Set.

Contains R52 060 Tank loco in black; an R18 cable drum wagon in black and grey; brown and blue; an R143 mineral wagon in black and grey; an R12 Shell BP petrol tank wagon in silver, black and red; an R16 ER brake van in fawn, black and white; a connecting clip and straight, curved and uncoupling track to form an oval 99 x 81 cms. Price 71/6d. Issued 1968. Deleted 1973.

£30

RS.26. Goods Train Set (BR)

Contains 060 class 3F loco and R33 tender in black; an R243 mineral wagon in off-white; an R219 bogey brick wagon in black and fawn; an R122 cattle wagon in fawn and white; an R211 Shell oil tank wagon in yellow and black; an R124 W.R. brake van in fawn and grey; complete with large circle track 129 x 81 cms. Price 88/9d. Issued 1962. Deleted 1969.

£35

RS.27. Diesel Rail Car Set.

Contains an R157 diesel power car in lime green with yellow lines and silver grey roof and a matching R158 diesel trailer car with built in seat unit, complete with super four track making an oval of 99 x 81 cms. Price 74/6d. Issued 1961. Deleted 1969.   £25

RS.28. 'Lord of the Isles' Set.

Contains R354S Dean single locomotive 'Lord of the Isles' in green, red and black with matching R37 tender. Coaches are in chocolate brown, fawn and white with black undercarriage. (G.W.R. livery). 2 R332 second class coaches and an R333 brake third coach; an R61 signal box, two R142H junction signals and an R495 level crossing together with some super track plus a power connecting clip, an uncoupling ramp, a left hand point and a right hand point to form an oval layout with a passing loop, 147 x 112 cms. Price 177/6d. Issued 1960. Deleted 1969.   £75

RS.29. Local Passenger Set.

Contains R150 460 class B12 loco with crew and an R39 tender in matching black and red line livery; an R626 BT crimson and cream composite coach and matching R627 BR brake second coach, complete with a section of straight, curved and uncoupling track to form an oval layout 99 x 81 cms. Price 79/6d. Issued 1963. Deleted 1969.   £40

RS.30. Crash Train Set.

Contains an R52 tank locomotive in black with emblems and decals, an R17 bolster wagon and matching truck for R127 operating crane truck, in black and rust; an R620 engineering department coach in black and grey, plus straights, double curves, power connecting clip and uncoupling ramp to form an oval layout 99 x 81 cms. Price 75/-. Issued 1964. Deleted 1969.   £30

RS.31. Diesel Freight Set.

Contains an R353 yard switcher; an R116 gondola; an R117 oil tanker; an R115 caboose. Liveries may vary. Together with super track, straight and double curves with power connecting clip and uncoupling ramp to form an oval 117 x 81 cms. Price 78/6d. Issued 1962. Deleted 1969.   £35

RS.32. Diesel Freight Set.

Contains an R155 diesel switcher in yellow, red and black; an R235 pulp wood car in grey and black; an R139 pick-up car in black, red and grey; an R115 caboose in green or red with a selection of super four track, straights and double curves, with power connecting clip and uncoupling ramp, forming an oval 117 x 81 cms. Price 101/-. Issued 1962. Deleted 1969.   £40

RS.33. Express Passenger Set.

In matching livery of dark orange, silver grey and black, with Intercontinental decals, consisting of an R55 diesel loco; an R442 baggage and kitchen car; an R443 diner; an R440 coach, an R441 observation car together with super track four, straights and double curves including a power connecting clip and uncoupling ramp forming an oval 142 x 97 cms. Price 126/3d. Issued 1962. Deleted 1969.   £50

RS.34. Blue Streak with Working Headlight

Beautiful set in matching purple and grey with yellow trim, comprising of an R159 double ended diesel loco; an R446 baggage

and kitchen car; an R447 diner, an R444 coach; an R445
observation car, together with super four track, straights and double
curves, with power connecting clip and uncoupling ramp forming
an oval 142 x 97 cms. Price 124/6d. Issued 1962. Deleted 1969.   £60

RS.35. The Indian Freighter Set.

Contains the famous 'Hiawatha' engine with smoke, in black, and
white or grey and white; an R32 matching tender; R26 stock car;
an R129 refrigerator car; an R136 box car; an R115 caboose.
Liveries may vary. Complete with super four track, straights and
double curves with power connecting clip and uncoupling ramp
forming an oval 142 x 97 cms. Price 139/6d. Issued 1962. Deleted
1969.   £50

RS.36. The Highwayman Set.

Contains a Co-Co EM2 loco in green, grey and black, plus twin
operating pantographs powered either from the overhead power
supply system or from the track itself; a BR maroon composite
coach and matching B.R. maroon brake second coach, complete
with super straight and curved track forming an oval 142 x 97 cms.
Please note the catenary system enables two trains to be operated
by separate power controllers simultaneously and independently on
the same track. Price 135/-. Issued 1965. Deleted 1970.   £40

RS.37. Frontiersman Set.

Now to one of the most sought after sets in the Triang range. The
famous Davy Crockett locomotive in red, black and golden yellow
with matching tender, with two old time coaches in golden yellow,
black and brown, complete with crew and smoke, straights and
double curves, power connecting clip and uncoupling ramp forming
an oval layout 117 x 81 cms. Price 102/6d. Issued 1963. Deleted
1970.   £50

RS.38. Snow Rescue Set.

Contains R138 snow plough in green, black and silver grey; an
R353 yard switcher in red and black; an R128 operating helicopter
car in green, red and black complete with helicopter; an R248
ambulance car in green, black and grey with red crosses on sides
and roof. The helicopter really flies. A number of straights and
double curves, power connecting clip and uncoupling ramp to form
an oval 117 x 81 cms. Price 87/6d. Issued 1964. Deleted 1970.   £40

RS.39. Goods Set.

Contains three trucks, a milk wagon and a guards van with a black
loco and matching tender; wagons can be in various liveries. One of
the best clockwork sets. Complete with an oval layout 117 x
81 cms. Issued 1964. Deleted 1969.   £35

MODEL	MB	MU	GC

RS.40. Clockwork Passenger Set.
Green and gold loco and tender and two coaches in maroon and silver. Complete with a section of curved and straight track to form an oval 91 x 71 cms. Price 25/-. Issued 1963. Deleted 1969. £35

RS.41. Passenger Clockwork Set.
Green, black and gold loco and matching tender with two carriages and guards van in silver grey. Complete with matching guards van and circular track to make an oval 117 x 81 cms. Price 58/3d. Issued 1960. Deleted 1969. £45

RS.42. Electric Goods Set.
Red and black loco, three coal trucks in grey and white and matching guards van; complete with a section of straight and curved rails, power connecting clip and uncoupling ramp to form an oval 91 x 71 cms. Price 53/3d. Issued 1959. Deleted 1969. £40

RS.43. Clockwork Goods Set.
Contains 557 diesel locomotive in purple with red wheels, an R10 open wagon in green and black; an R16 E.R. brake van in black, chocolate and grey with a track forming a circular 81 cms. Price 19/11d. Issued 1964. Deleted 1970. £25

RS.44. Picador Set.
Contains black loco, a green container truck, a purple container truck and a black and brown wagon; complete with a circular track 91 x 71 cms. Price 95/-. Issued 1964. Deleted 1969. £35

RS.45. The Mail Man Set.
Contains green locomotive with operating pantographs. This model can be worked from a high-low dual control system. With maroon and silver coach and maroon and silver grey mail coach, complete with straight and curved track, power connecting clip and uncoupling ramp. Price 119/6d. Issued 1964. Deleted 1969. £35

RS.46. Intercity Continental Set.
Contains a silver grey streamlined loco and matching silver and black continental coaches and restaurant car, complete with straight and curved rails, power connecting clip and uncoupling ramp. Issued in limited edition in 1964. Deleted 1966. £50

RS.47. Double Train Set.
Double set contains red and black loco, a grey and black open wagon, a green, grey and black cattle truck and a brown, black and white guards can, plus a green and black loco with matching tender and two coaches in matching maroon, golden yellow and grey, with a large section of straight and curved track to satisfy any collector. In special presentation box, complete with power clips and uncoupling ramps to form a large oval approximately 168 x 95 cms. Price 168/-. Issued 1964. Deleted 1966. £75

RS.48. Victorian Train Set.
Loco and tender in green, grey and brown livery with two old style Victorian coaches in matching chocolate and cream with black wheels and grey or silver roof. Complete with crew, curved and straight rails, power connecting clip and uncoupling ramp to form an oval 117 x 81 cms. Price 85/-. Issued 1964. Deleted 1969. £50

RS.48/A. Ferry Train Set.

Made in limited edition with grey, black and red loco and matching tender and four goods wagon s in various liveries with guards van; complete with crew, straight and curved track forming an oval 117 x 81 cms. Power connecting clip and uncoupling ramp in special presentation box with dockland scenery. Price 105/-. Issued 1964. Deleted 1966.   £50

RS.49. Docklander

Another set made in limited edition with famous 'Connie' loco in yellow and black. With white refrigerator van, Shell tanker van, two grey coal wagons and a guards van in cream and white; complete with straight and curved track, power connecting clip and uncoupling ramp, to form an oval 117 x 81 cms. Price 99/-. Issued 1964. Deleted 1966.   £50

RS.50. Defender Set.

Contains an R152 diesel shunter in green and black; an R341 searchlight wagon in green and black with a searchlight that actually shines; an R343 rocket launcher in green and black; an R249 exploding car in red and black; complete with double curves, straights, power connecting clip and uncoupling ramp. Track forms an overall layout 117 x 81 cms. In a special presentation box. Price 109/6d. Issued 1964. Deleted 1969.   £50

RS.51. Freightmaster Set.

Contains green, yellow and black diesel loco with guards van and three open wagons in matching brown, black and grey livery, with straights and double curves, power connecting clip and uncoupling ramp to form a layout 117 x 81 cms. Price 65/-. Issued 1964. Deleted 1969.   £35

RS.51/A. The Big Freightmaster Set.

Contains an A1A-A1A diesel locomotive in green, black and grey with white lines and yellow and red design; an R123 horse-box in brown, black and grey; an R122 cattle wagon in brown, black and grey; an R15 milk tank wagon in white and black with 'U.D.' decals; an R561 container wagon in purple, brown and black with 'Triang' decals; an R340 container wagon in grey, brown and black; an R113 dropsides wagon in chocolate and black; an R16 E.R. brake van in brown, black and grey; complete with straights and double curves, power connecting clip and uncoupling ramp to form an oval 117 x 81 cms. Price 99/6d. Issued 1964. Deleted 1970.   £50

RS.52. The Blue Pullman Set.

In matching blue, white, black and red with grey roof. This beautiful set consists of an R555 diesel Pullman motor car; an R426 Pullman parlour car; an R556 non-powered Pullman car, complete with straights, double curves, power connecting clip and uncoupling ramp to form an oval 117 x 81 cms. In special presentation box. Price 97/6d. Issued 1964. Deleted 1969.   £45

RS.53. The Caledonian Set.

In authentic Caledonian livery with loco, matching tender, two coaches, sleeping car and restaurant car and guards van. Complete with a large amount of track, power connecting clip and uncoupling ramp to form layout 117 x 81 cms. Price 100/-. Issued 1963. Deleted 1969.   £50

RS.54. The Flying Scotsman Set.

In LNER livery consisting of loco, coaches, sleeping car, restaurant car and guards van with a large track layout approx 117 x 81 cms. Price 105/-. Issued 1964. Deleted 1970. £50

RS.59. Electric Passenger Set.

Contains R659 black tank locomotive; an R720 coach in maroon, lemon and grey with black chassis and wheels; with curves, power connecting clip forming a layout 81 cms in diameter. Price 39/6d. Issued 1964. Deleted 1969. £30

RS.61. Old Smokey Train Set.

Consisting of the Grimy in black or the rare very dark brown, which is worth treble, complete with tender and crew. Includes a coach in maroon, lemon and grey and a utility van in green, red, black and grey with twelve opening doors; an oval of super four track and a ramp for automatic coupling to form an oval approx 100 x 80 cms. Price 69/6d. Issued 1965. Deleted 1973. £35

RS.62. Car-A-Belle Set.

Contains black loco and two car transporters in grey and black, complete with twelve cars, an E.R. brake van in brown, black and grey, complete with super track four and automatic uncoupling ramp to fom an oval 117 x 81 cms. Price 97/6d. Issued 1965. Deleted 1969. £50

RS.70. Clockwork Goods Set.

Green and black loco with three trucks and a guards van. Liveries could vary. With an oval track approximately 81 cms in diameter. Price 20/6d. Issued 1965. Deleted 1970. £25

RS.85. Clockwork Set.

Green loco with matching tender with black design and two carriages. With circular track approximately 81 cms. in diameter. Price 35/11d. Issued 1968. Deleted 1973. £25

RS.86. Clockwork Goods Set.

Black loco with matching tender and four goods vans in various liveries. With an oval track 81 cms. in diameter. Price 35/11d. Issued 1968. Deleted 1970. £25

RS.89. Rail Freight Set.

Contains a diesel shunting loco in blue, black and yellow; a bogey flat car in fawn and black with three Minic vans in red, white and blue; plus a bogey flat car in fawn and black with two 20 ft. freight liner contains in maroon and grey, complete with 'Freight Liner' decals. Set contains a section of oval track with automatic coupler to make a circle approx. 81 cms. in diameter. Price 95/-. Issued 1969. Deleted 1973. £40

RS.90. The Albert Hall Pullman Set.

A rare and famous set containing the Albert Hall express locomotive in green, black and gold lines with matching tender, plus three Pullman coaches in matching fawn, cream, black and white. Complete with a large oval track, level crossing and automatic uncoupler to make an oval of 127 x 91 cms. Set also has special presentation box. Price 125/-. Issued 1968. Deleted 1970. £75

RS.91. The Caledonian Special.

Blue, black and grey livery with matching tender and a Caledonian brake compo coach and a first and third compo coach in maroon, white and black, with a large section of track, uncoupling ramp and connecting point. This famous loco used to run between Carlisle and Glasgow during early part of 20C. Layout made an oval track approx 127 in x 91 cms. and came in a special presentation box. Price 100/-. Issued 1965. Deleted 1970.

£75

RS.1/1969. Clockwork Steam Set.

These simpler life like toy train sets were designed for the young newcomer to model railways. However, they were not the success that had been predicted and were not as reliable or successful as the earlier sets. First set had two keys, a steam locomotive in red; an open wagon in green; a flat wagon in red, although liveries could vary. A circle of track approx. 81 cms. Price 50/-. Issued 1969. Deleted 1973.

£15

RS.2/1969. Clockwork Steam Passenger Set.

Steam loco in red with two keys and two coaches in yellow and brown with a circle of track 81 cms. in diameter. Price 50/-. Issued 1969. Deleted 1973.

£20

RS.3/1969. Large Clockwork Steam Freight Set.

A red loco with two keys, a green flat wagon and a blue flat wagon; an open red wagon and a yellow tank wagon. The flat wagons contained a red Minic car and a silver Minic van. Complete with an oval track approx 100 cms. in diameter. Also a level crossing. Price 65/-. Issued 1969. Deleted 1973.

£25

RS.4/1969. Electric Passenger Train Set.

Contains a blue locomotive and two coaches in yellow and brown. With a battery box controller and a circle of track approx. 81 cms. Price 60/-. Issued 1969. Deleted 1973.

£15

RS.5/1969. Electric or Diesel Set.

Contains a blue diesel locomotive; an open wagon in red; a tank wagon in yellow; a flat wagon in green containing a silver Minic van; a battery box controller and a circle of track approx. 100 cms. in diameter. Price 65/-. Issued 1969. Deleted 1973.

£20

RS.6/1969. Large Electric Freight Train Set.

Contains blue diesel loco, a green flat wagon containing a red Minic car; a blue flat wagon containing a silver Minic van; an open wagon in red and a tank wagon in orange. Complete with a battery box controller and an oval of track with a level crossing approx. 100 cms. in diameter. Price 72/-. Issued 1969. Deleted 1973.

£25

COMBINATION SETS

RM.A. Railway Motorway Combination Set.

These were sets specially designed for the Triang Minic Motorway and Railway enthusiasts and the first contained a black or grey loco with matching tender; two maroon, black and grey coaches; a black or grey Minic car and a red Minic car. Complete with crew and track etc. to make an oval 168 x 95 cms. Price 219/6d. Issued 1964. Deleted 1970.

£50

RM.B. Connie Set.

Contains R355Y 'Connie' loco in yellow and black with red decals; an R.10 open wagon in grey; R14 fish van in white, grey and black; R.16 E.R. brake van in brown, black and white; a power clip; uncoupling ramp; inclined piers set; complete with straight and curved track to form a large circle approximately 150 x 100 cms. Contains also M1573 Aston Martin; M1574 Porsche; a level crossing with speed and direction controllers. Loco and cars operate from a 12 volt DC system. Price 170/-. Issued 1964. Deleted 1970.

£50

RM.C. 3F Class Loco Set.

Contains a black loco and tender; R339 sleeping car in maroon, black and grey; M1541 Rolls Royce in black; M1543 Humber Super Snipe in red; RM922 railway car transporter, complete with speed and direction controllers, power clips, uncoupling ramps and a large section of track forming a layout 152 x 81 cms. Price 199/6d. Issued 1964. Deleted 1970.

£60

RM.D. Motor Rail Set.

In beautiful presentation box containing locomotive in red and green; a car controller; car loading ramp and transporter; an oval of roadway with junction and circle of railway track with points to form a layout of 152 x 81 cms. Price 175/6d. Issued 1968. Deleted 1973.

£50

R17C. Flat Wagon with Minic Load.

This set contains car transporter in light fawn with six cars in various liveries; a flat truck in black and brown with a green Minic car and a loader truck in black and maroon with decals in white. Price 42/-. Issued 1966. Deleted 1970.

£30

R128. Helicopter Car Set.

Contains trackside trip lever to clip fit to track, complete with helicopter in red, blue and black, 10.8 cms. long approx and a helicopter wagon in green 8.3 cms long. approx. As helicopter car passes the trip lever the 'chopper' is automatically launched. Price 14/6d. Issued 1963. Deleted 1970.

£15

R186. Field Fencing Set.

White or grey consisting of seven fences; a five bar gate; a gate base; a stile and six fence bases. Price 2/6d. Issued 1963. Deleted 1970.

£1

R265. Station Lamps Set.

Lime green, black and silver. Set of station lamps which fit to platform sections or stand in goods yards adding brilliance to the night scene of any layout. 15 volts AC. Price 22/6d. Issued 1962. Deleted 1969.

£2

MODEL	MB	MU	GC

R266. Station Lamps.
In lime green, silver and black. Issued 1963. Deleted in 1970. 15 volts AC. Price 12/-.

£2

R268. Bell Signal Set.
Contains two block instrument units each consisting of visual indicator, bell and tapping key, together with interconnecting wire. 15 volts AC or 12 volts DC which enables a train to be passed from one track section to another using authentic railway bell signalling procedure. Price 45/-. Issued 1964. Deleted 1970.

£6

R298. Home Maintenance Set.
In special presentation box containing senior tool kit in red bag, a track tester, three screw drivers, a spike, commutator cleaning knife and tweezers and four drums of wire in purple, red, green and black. Twelve fish plates, six contact plugs, six tag washers, six couplings, four motor brushes and one packet of track pins with instruction leaflets. Price 17/6d. Issued 1964. Deleted 1970.

£10

R319. Conversion Kit Set.
A conversion kit for use with the R407 hand-operated turn table. It cleverly converts the R407 to R408 electrically operated 12 volts DC. Price 29/6d. Issued 1964. Deleted 1970.

£5

R346. Rocket Set.
One of the best investments in this series. A model of the famous Stephenson's Rocket train, fully detailed in yellow, black and red with crew and silver funnel. It was worked with moving connecting rods. Includes a tender and one coach, with a selection of track to make an oval of 81 cms. Special presentation box. Price 47/6d. Issued 1965. Deleted 1970.

£50

R346 A. Large Rocket Set.
Same livery as above but this time with three carriages and a large section of track to make an oval of 127 cms x approx. 91 cms. Set was made for the 1966 Toy Exhibition of London. Price 75/-. Issued 1966. Deleted same year.

£100

R386. Assembly Pack 1.
Contains locomotive 'Princess Elizabeth' with crew and tender in authentic green, grey and black livery, complete with assembly pack. No previous experience necessary to complete this beautiful model. Price 50/-. Issued 1968. Deleted 1973.

£10

R400. Transcontinental Mail Coach Set.
Red, black and grey, complete with pick-up hook section, operating ramp, receiving bin and mail bags. Has action accessory which can be used with either Super 4 or Series 3 track. With 'Transcontinental' emblems and wording in gold. 26.2 cms approx. Price 24/6d. Issued 1962. Deleted 1970.

£20

R401. Transcontinental Mail Coach Set.
Blue, grey and black with gold line and 'Transcontinental' wording and emblems, otherwise as R400. 25.1 cms. approx. Price 24/6d. Issued 1962. Deleted 1970.

£25

R402. Operating Royal Mail Coach Set.
Contains Travelling Royal Mail Coach in deep maroon with 'Royal Mail' in gold lettering, pick-up hook, receiving bin, operating

ramps and mail bags. The hook, bin and ramps clip-fit to straight
track sections with automatic operation. 26.2 cms approx. Price
24/6d. Issued 1964. Deleted 1970.   £18

R403. Ore Wagon Set.

Red and black, identical with R404, only with an R214 operating
ore wagon in place of R111 operating hopper car. 14.4 cms approx.
Price 17/6d. Issued 1962. Deleted 1970.   £10

R404. Hopper Car Set.

Red and black, consisting of R111 operating hopper car, hopper
unloading bridge, a set of R85 arched piers, two 453 high level
piers, a brush and shovel and mineral bunker with mineral.
14.4 cms. Price 19/11d. Issued 1962. Deleted 1970.   £15

RT.405. Colour Light Signal Set.

Yellow, black and maroon, electrically operated from 15 volts AC.
Changes colour on operation of level frame section and can be
connected up to R406 automatic train control set. Price 15/6d.
Issued 1962. Deleted 1970.   £6

R406. Automatic Train Control Set.

Brown, black and silver, containing relay, two actuators, two R497
isolating tracks and connecting leads. One train can be used to start
and stop another, or to control colour light signals whilst travelling
round the layout, all completely automatic with no one at the
control. Price 17/6d. Issued 1962. Deleted 1970.   £6

R416. Catenary Set (450 cms).

Green and black, containing sufficient equipment to fit overhead
power supply to the following train sets: RS22, RS23, RS34, RS35,
and RNC. Price 29/6d. Issued 1964. Deleted 1970.   £5

R417. Catenary Set (300 cms).

Black, green and white, containing sufficient equipment to fit
overhead power supply to the following train sets: RS24, RS29,
RS30, RS37, RS38, RS44, RS45, RS48, RS50, RS51, RS52, RMA
and RMC. Price 22/6d. Issued 1964. Deleted 1970.   £5

R418. Catenary Extension Set.

Green, white and black, containing sufficient equipment to fit
overhead power supply to 6 ft. of track and above a diamond
crossing and two points. Price 14/11d. Issued 1964. Deleted 1970.   £3

R429. Twin Line Conversion Set.

Contains two points, ten straights, eight curves, an L/RAD and
power clip. Price 42/-. Issued 1966. Deleted 1970.   £5

R432. Girder Bridge Set.

Grey and brown, consisting of R77 bridge support, R78 girder
bridge, two R453 high level piers, two R457 inclined piers, two
R456 straight side walls and one or two gradient posts. Price 37/6d.
Issued 1963. Deleted 1970.   £5

R439. Passing Loop Set.

Contains two points, eight straights, two curves and a left/RAD.
Price 25/6d. Issued 1966. Deleted 1970.   £5

R454. Siding Set.

Contains one point, three straights, a curve L/RAD, uncoupling, buffer stop and loading gauge. Price 14/11d. Issued 1966. Deleted 1970. — £3

R458. Small Station Set.

Blue/grey and white, black and maroon. Colours may vary with the possible use of yellow. Consists of two R460 platform units, a straight, two R464 platform ramps, a double curve, R473 ticket office with built in lighting unit and one R465 steps unit. Price 23/-. Issued 1962. Deleted 1970. — £7

R459. Large Station Set.

In white, blue/grey, red or maroon, yellow, white and black. This set consists of three R460 platform units, a straight, two R464 platform ramps, two double curves, R465 steps units, R466 straight canopy set, two R469 seat units, two R470 name boards complete with selection of station names, two R471 platform fencing units and one R473 ticket office with built-in lighting units. Price 36/-. Issued 1962. Deleted 1970. — £10

R588. Island Platform Set.

Yellow, maroon, white, grey and blue, consisting of platform, canopy, seats, ticket office and newstand. 59 cms approx. Price 16/-. Issued 1965. Deleted 1970. — £6

R589. Ultra Modern Station Construction Set.

This magnificent set has the varied colours of white, green, blue, brown, black and grey, plus many varied colours, including green and yellow. Contains well over 550 parts to build an over-the-line station based on real principles. Fit together rolled steel joists, no adhesives required, then add cladding. All parts are pre-coloured and set can be taken down and used to construct other buildings. In special presentation box. Priced 49/6d. Issued 1966. Deleted 1970. — £25

R644. Inter-City Train Pack.

Blue, yellow, black and grey, containing a Bo-Bo electric locomotive, two second-class coaches and a first-class brake coach, with full Inter-city emblems, decals and numbers. Price £5.19.6d. Issued 1968. Deleted 1973. — £20

R644A. Inter-City Train Pack.

Turquoise, black, yellow and blue, containing a Bo-Bo electric locomotive, two second-class coaches, and a brake first-class coach with the new B.R. passenger stop, complete with interior lighting. Price £5.19.6d. Issued 1969. Deleted 1973. — £25

R645. Freightliner Train Pack.

With blue, orange-yellow and black Hymeck diesel locomotive, and three grey, maroon, black and white Freightliner wagons. With authentic B.R. Emblems and wording, numbers, etc. Price £5.17-.d. Issued 1968. Deleted 1973. — £15

R672. Honest John Missile Pad.

Grey, fawn, black and red. Hand-based missile site for home defence. Missile has rubber warhead and launcher has powerful spring with variable elevation hand trigger. Price 17/6d. Issued 1968. Deleted 1973. — £5

FIGURE SETS

Five Train Figures.

Contains five figures in authentic blue BR livery, two guards, a
driver, fireman and motorman. Price 2/6d. Issued 1961. Deleted
1970.   £2

R283. Platform Figures.

Contains guard in grey and white with flag, a porter in grey and
white carrying two brown cases, a city gent in grey and white with
black umbrella and newspaper, a sitting woman in green with
yellow hat and yellow gloves and a sitting man in brown with
purple cap. Price 2/6d. Issued 1961. Deleted 1970.   £2

R284. Five Coach Figures.

Contains attendant in grey and white, carrying drinks on tray, two
man tops in brown and two woman tops in blue. Price 2/6d. Issued
1961. Deleted 1970.   £2

RML70. Set No. 1 (Pedestrian).

Contains seven figures, man in grey and brown, running with
newspaper, man in blue walking, lady in grey skirt and purple
jumper holding hand of small girl in green dress, lady in orange
and green carrying shopping bag, gentleman in light suit reading a
newspaper and carrying an umbrella and gentleman walking in grey
suit walking with brown brief-case. Price 3/6d. Issued 1964.
Deleted 1970.   £3

RN71. Set No. 2. (Workmen).

Contains two painters in white overalls, milk man in blue and
white carrying milk crate, man in green with a brown shovel,
postman in blue with a bag of mail and lady in white overall. Price
3/-. Issued 1964. Deleted 1970.   £3

RML72. Set No. 3 (Children).

Contains a boy in green running with school bag; two boys playing
leap frog, one in green and one in blue; a girl in a blue dress
playing with a hoop; a girl in a green dress skipping and a girl in a
red dress standing. Price 2/9d. Issued 1964. Deleted 1970.   £3

RML73. Set No. 4. (Urban Figures).

Contains a porter carrying green bag, a painter or window cleaner
with ladder and bucket, a station attendant with hand board, a
policeman on point duty and a policeman approaching a man in
brown, carrying a yellow sack. Price 3/-. Issued 1964. Deleted 1970.   £4

RML74. Set No. 5. (Industrial Workers).

Contains a workman in blue overalls with shovel, a newsboy in
green with papers, a mechanic with tool box and spanner, a driver
in blue and two workers, one in brown and one in blue. Price 3/-.
Issued 1964. Deleted 1970.   £3

RML75. Set No. 6. (Road Workmen).

Contains road sweeper in brown and green, chimney sweep in
brown and blue carrying brushes, man in blue and brown carrying
bucket of cement, man in green and purple with drills, man in grey
and yellow digging with shovel and a foreman inspecting a sign.
Price 3/-. Issued 1964. Deleted 1970.   £3

R164. Battle Space Commandos.

Brown and white. Specially designed to man battle space units. All
six figures are armed and in various positions. Price 2/6d. Issued
1966. Deleted 1970.   £3

R165. Stephenson's Rocket Crew.

Consists of driver in blue and fireman in green, white and brown
with grey cap. Specially made to go with the famous Stephenson's
Rocket Set. Price 10d. Issued 1968. Deleted 1973.   £3

R413. Locomotive Crew.

Contains fireman with shovel and engine driver in authentic BR
colours. Price 9d. Issued 1964. Deleted 1970.   £1

MODEL-LAND

Owing to the tremendous success of the Triang Railway locomotive sales, the company decided to introduce a series in 1964 which they called 'Model-Land'. It was with every kit minutely detailed, simple but strong and beautifully coloured to eliminate the difficul task of painting the finished model. This meant that children of a much younger age than normal could partake in this new edition of items to bring more pleasure and excitemen to the already existing model railways of perfection. Lighting units and working models operated either from batteries or any 9/12 volts DC supply. All the buildings were o British style and very carefully proportioned to blend with 00/HO and TT layouts. They have proved to be a very good investment, especially for people who specialized in building fine layouts to enter into the various exhibitions and competitions available to the collectors of Triang Railways.

MODEL	MB	MU	GC
RML1. Village Inn. Red, white, black and yellow, complete with fawn base and village sign in white, red and black. Price 7/6d. Issued 1964. Deleted 1969.	£10	£7	£5
RML2. The Grange. A fine establishment in black and white, complete with fawn base and potted plants. Price 6/11d. Issued 1964. Deleted 1969.	£12	£9	£7
RML3. Wood Shed. Dark brown and red, with grey water tank and white boiler chimney. Price 2/11d. Issued 1964. Deleted 1969.	£3	£2	£1
RML4. Marigold Cottage. Red and white with rust and grey roofs. Price 6/6d. Issued 1964. Deleted 1969.	£9	£7	£5
RML5. Dove Cottage. Red, white and green with fawn base. Price 4/11d. Issued 1964. Deleted 1969.	£7	£6	£4
RML6. Hardware Shop. Cream, green, black and white with white fencing and outer red and cream brick wall. Price 9/6d. Issued 1964. Deleted 1969.	£10	£8	£6
RML7. Garage. Red and white with yellow gate and red petrol pump on black base. Price 9/6d. Issued 1964. Deleted 1969.	£15	£13	£10
RML8. Accessories. Consists of horse trough, 'Inn' sign & war memorial, country stile and village stocks. Colours were white, black, brown and fawn. Price 3/6d. Issued 1964. Deleted 1969.	£4	£3	£2
RML9. Oak Tree Cottage. White walls with blue tint, red windows and doors and a fawn roof, complete with green plant decor on fawn base. Price 6/6d. Issued 1964. Deleted 1969.	£10	£7	£5
RML10. Hollywood Bungalow. This extra-special model has white walls and a green roof, brown oak doors, fawn outer wall with black gate and a green and pink garage with green and flower decor. Price 12/6d. Issued 1964. Deleted 1969.	£15	£12	£10

RML11. San Fernando Bungalow.
White, black, green and fawn on silver-grey base. Price 7/11d.
Issued 1964. Deleted 1969.

| | £12 | £10 | £8 |

RML12. Bermuda Bungalow.
White with green windows, blue roof, plant decor, brown bird
house. On silver/grey base. Price 8/11d. Issued 1964. Deleted 1969.

| | £15 | £13 | £10 |

RML13. Kent Bungalow.
Pink with blue roof, dark oak windows, yellow door and white
chimney on brown base. Price 6/6d. Issued 1964. Deleted 1969.

| | £9 | £7 | £5 |

RML14. Post Office.
White and dark brown with green windows and red doors. This is
combined post office and greengrocers. Complete with decals etc.
On fawn base. Price 7/11d. Issued 1964. Deleted 1969.

| | £15 | £13 | £10 |

RML15. Ye Olde Tea Shoppe.
White, green and black with red doors and yellow and red sign. On
fawn base. Price 6/11d. Issued 1964. Deleted 1969.

| | £15 | £13 | £10 |

RML16. Villa Capri.
White with blue doors and windows and a single yellow door,
brown roof and white fence veranda, with flower and plant decals.
Price 7/6d. Issued 1964. Deleted 1969.

| | £15 | £13 | £10 |

RML17. Village Church with Chimes.
This beautiful building h as off-pink walls, a red rust roof, fawn
doors, a flag with a red cross on a battlement-shaped upper roof.
Complete with wall clock and chimes. On black base. Price 24/-d.
Issued 1964. Deleted 1969.

| | £25 | £20 | £15 |

RML18. Heathview.
White, with yellow garage door and windows and brown roof.
Complete with blue door and plant decals. On brown base. Price
/6d. Issued 1964. Deleted 1969.

| | £12 | £10 | £8 |

RML19. Parkview.
Yellow, white and brown with plant and flower decals. On fawn
base. Price 8/6d. Issued 1964. Deleted 1969.

| | £12 | £10 | £8 |

RML20. Wishing Well.
Silver/grey and brown on a mustard base. Price 2/11d. Issued 1964.
Deleted 1969.

| | £3 | £2 | £1 |

RML21. Semi-Built Bermuda Bungalow.

Purple, white and woodgrain, showing clearly the unfinished roof
and interior. On a silver and black base. Price 7/11d. Issued 1964.
Deleted 1969. — £9 — £7 — £5

RML22. Lighting Unit.

Black and yellow. 12 volts AC/DC. Price 2/9d. Issued 1964.
Deleted 1969. — £4 — £3 — £2

RML23. Pit Head Gear.

Silver/grey, orange and white on dark brown base. A fine model of
an actual pit head, which could be seen around the mining villages
for many years. Price 10/6d. Issued 1964. Deleted 1969. — £12 — £10 — £8

RML24. Pit Head Canteen.

Dark brown and red, on black base. Price 10/6d. Issued 1964.
Deleted 1969. — £10 — £8 — £6

RML25. Winding House.

Red, green, yellow and black. To go with RML23 pit head gear.
Price 12/6d. Issued 1964. Deleted 1969. — £14 — £12 — £9

ML26. Winding Engine.

Red, silver and black, made to use with RML23 pit head gear to fit
inside RML25 winding house. 9/12 Volts D.C. Price 27/6d. Issued
1964. Deleted 1969. — £25 — £20 — £15

RML27. Boiler House and Chimney.

Red, black, yellow and silver. Price 30/-. Issued 1964. Deleted
1969. — £20 — £15 — £10

RML35. Gas Holder.

White and brown. Price 12/6d. Issued 1964. Deleted 1969. — £15 — £10 — £8

RML44. Small Shop With Old Style Office Block.

Yellow, red and dark cream with 'Triang Toys' sign and shop
decal. Price 9/6d. Issued 1964. Deleted 1969. — £15 — £10 — £8

RML45. Medium Shop with Old Style Office Block.

Cream, lemon, red, green and rust red. Price 14/6d. Issued 1964.
Deleted 1969. — £18 — £15 — £12

RML46. Multiple Store with Modern Office Block.

Bright cream, lemon, red, white and green, with full window store
decals, signs etc. Price 21/6d. Issued 1964. Deleted 1969. — £25 — £20 — £15

RML52. De-Havilland Trident Airliner.
Blue, grey and off-white, red and Union Jack decals. 'BEA' in large white letters on red background on sides and wings. A rare find in the Triang range and one of the best investments. With powered remote-controlled taxiing and strong steering. 9/12 Volts D.C. Price 52/-. Issued 1964. Deleted 1966.

£75   £60   £50

RML55. Police Station.
Red, cream and blue, with grey roof. Green door and drainpipes. Price 21/-. Issued 1964. Deleted 1969.

£25   £20   £15

RML56. Town Hall.
Red and cream with green or grey roof, green drain-pipes and doors. Complete with clock-tower and Union Jack. Price 30/-. Issued 1964. Deleted 1969.

£35   £30   £25

RML57. Georgian House.
Cream, yellow, with green or grey windows, green roof and drain-pipes. Oak door and window-boxes. Price 21/-. Issued 1964. Deleted 1969.

£35   £30   £25

RML58. Village Church Without Chimes.
Identical to RML17, apart from there being no chimes. Price 52/6d. Issued 1964. Deleted 1969.

£20   £15   £10

RML59. A Large Factory.
Cream, grey and black, complete with large black chimney. Price 59/6d. Issued 1964. Deleted 1969.

£25   £20   £15

RML60. Supermarket.
Dark red, black, green and silver/grey with slate grey base. Price 51/-. Issued 1964. Deleted 1969.

£30   £25   £20

RML61. Set of 3 Pylons.
Black and silver/grey with green bases. Price 14/-. Issued 1964. Deleted 1969.

£12   £9   £5

RML62. Spring Green.
Landscape material. Price 1/3d. Issued 1964. Deleted 1969.

£3   £2   £1

RML63. Autumn Green.
Landscape material. Price 1/3d. Issued 1964. Deleted 1969.

£3   £2   £1

RML64. Landscape material. Price 1/3d. Issued 1964. Deleted 1969.

£3   £2   £1

RML65. Bus Shelter.
Green and cream, with black seat. Price 3/6d. Issued 1964. Deleted 1969.

£8   £6   £3

RML66. Village School.
Rust red, cream or off-white, black, grey and green. Price 37/6d. Issued 1964. Deleted 1969.

£50   £45   £40

RML70. Set of Pedestrians.
Seven pedestrians, male and female in a variety of dress and colours. Price 2/6d. Issued 1964. Deleted 1969.

£3   £2   £1

MODEL	MB	MU	G
RML80. Taxi. Black, with taxi sign and white interior. Price 5/6d. Issued 1964. Deleted 1969.	£15	£10	£7
RML81. British Railways Van. Mustard, black and red, with 'B.R.' decals. Price 5/11d. Issued 1964. Deleted 1969.	£18	£15	£1
RML82. G.P.O. Van. Red, silver and black, with 'Royal Mail' decals on sides. Price 5/6d. Issued 1964. Deleted 1969.	£30	£25	£2
RML1801. Filling Station. Silver/grey, with a green on dark fawn base and 'Shell Filling Station' decals in black. Also set of pumps and lamps. Price 9/11d. Issued 1964. Deleted 1969.	£15	£12	£1
RML1802. Service Bay. Silver/grey, red, black and orange on dark fawn base. Price 12/6d. Issued 1964. Deleted 1969.	£15	£12	£10
RML1803. Bus Garage. Red and medium fawn with blue windows and 'London Transport' decals in black. Price 12/11d. Issued 1964. Deleted in 1969.	£20	£15	£1
RML1804. Fire Station. Light brown, red and black, on dark fawn base. Price 14/-. Issued 1964. Deleted 1969.	£15	£12	£10
RML1805. Bus Station. A unique and rare set. Contains ticket office, manager's office, bus station stands complete with name boards in red, black, grey and green, including two buses in red and black. Price 56/-. Issued 1964. Deleted 1969.	£75	£65	£50
RML1806. News-Stand. Red, black and silver/grey on dark fawn stand, complete with newsagent and news-boy with cycle. Very rare. Price 21/-. Issued 1964. Deleted 1969.	£75	£65	£50
RML1807. Bungalow with Automatic Garage and Smoking Chimney. White, red, blue and cream, complete with 12 volt D.C. and special track with trip lever to go with either railway or Minic motorways set. Price 26/6d. Issued 1964. Deleted 1969.	£30	£25	£20
RML1808. Bungalow With Garage and Non-Smoking Chimney. Red, white, and black with blue doors and light fawn base. Price 15/-. Issued 1964. Deleted 1969.	£12	£10	£8
RML1809. Goods Shed. Red, green and blue on grey base. Price 11/6d. Issued 1964. Deleted 1969.	£15	£12	£10
RML1810. Footbridge. Grey and black. Price 7/-. Issued 1964. Deleted 1969.	£5	£4	£3

RML1811. Bus Shelter.

Green and black, complete with timetable board. Price 5/11d.
Issued 1964. Deleted 1969.

£12 £10 £8

RML1812. Motorway Service Station and Bar with Cafe.

White, red and dark blue, complete with various flags of many nations, on light green base. Price 30/-. Issued 1964. Deleted 1969.

£50 £45 £40

RML1813. Motel Restaurant.

Black, white and blue with large green roof and 'Motel' sign in orange. Price 8/9d. Issued 1964. Deleted 1969.

£10 £8 £6

RML1814. Motel Chalet and Garage

White, dark yellow, medium green and black, on silver/grey base. Price 12/3d. Issued 1964. Deleted 1969.

£15 £12 £10

RML1815. Car Park.

Silver/grey, black and red on dark fawn base. Complete with two attendants in white uniforms and parking space for 30 Minic cars. Price 21/-. Issued 1964. Deleted 1969.

£15 £12 £10

RML2001. The Operating Big Wheel.

An authentic replica of the Big Wheel which can be seen on any fairground. The wheel in silver/grey with purple, red, green, yellow and blue bucket seats. Works with a 3 volt dry battery. Set on a dark red base with yellow control cabin and two attendants. Base is unique with surrounding white railways. A first class investment. Price 32/-. Issued 1964. Deleted 1969.

£75 £65 £50

RML2002. The Large Roundabout.

With the flick of a switch, this unique roundabout comes to life with the aid of a 3 volt dry battery. Orange, blue, black, red, white and green on light fawn base with green control cabin. The roundabout is complete with blue, white, black, yellow, red and silver horses. Price 35/-. Issued 1964. Deleted 1969.

£75 £65 £50

RML2003. Operating Octopus.

Minic Railways and Motorways bring the fantastic colour, music and all the fun of a fair with this traditional operating octopus. The bucket seats are in blue, red, green and yellow with silver stabiliser arms. On dark brown base with octopus decals in black and blue with yellow outlines and control box. Price 30/-. Issued 1964. Deleted 1969.

£50 £45 £40

RML2004. Set of Fairground Attendants.

Nine male and female figures in a variety of dress and uniform. Price 14/6d. Issued 1964. Deleted 1969.

£20 £15 £10

MINIC MOTORWAYS VEHICLES

INTRODUCTION

THE NEW TABLETOP FULL OF THRILLS

Minic-miniatured Hi-Speed electric motor racing was first introduced in 1960, although it took while to catch on. By the early part of 1963, the tabletops in the United Kingdom and other parts of the world began to fill with racing cars, buses, lorries, caravans, and a whole series of very colourful service stations, cottages, hotels, and scenery. It was all part of the dramatically re-created exciting world of Minic Motor Racing with fabulous GT cars! superbly engineered. There were miniaturised cars and track in compact scale designed to give children and grown-u alike many miles of track per square foot plus exciting new automated accessories and trackside decor exclusive to Lines Brothers. Minic Marvels brought alive in any household all the thrills full-sized motor racing.

Not only racing, but holiday scenes, city glamour and the views one would see in everyday li or while on holiday, while shopping, or when one went to an open air or indoor event. Police and crime, banks and security, road haulage and postal deliveries could be re-enacted in your living room, attic, or cellar. The tiny figures were life-like, high quality models with reliability and were safe for any child to play with. The makers had thought of everthing, and every product was a winner.

Model racing circuits were planned in association with 'MOTOR' the famous Motoring Magazine. The company were ahead in every direction. Production car racing, a spin along the Motorway or a long Road/Rail journey on a car transporter, the versatile Minic Motorways system served all aspects of miniature motoring.

The Minic Motorway system fits perfectly with most railways and rolling stock. You could r a lorry into a goods depot to pick up a freight container right from the rail sidings. A batch of new cars could be delivered by rail and then be driven off and on the train with the traffic kept waiting at the railroad crossing till the Express went through. Easy-to-assemble, clip together buildings added to the fun, all made by Minic. Layouts of the worlds top ten racing circuits we available, and frontier posts with life-like guards etc., plus well known villages which could be built in real-life-like form.

Any person lucky enough to have the full range of Minic Motorway models has a considerabl investment. The run of Minic Motorways ran from 1960 until 1970, although a few sets and models may have survived until about 1977. The only products available today in this discontinued series will be from collectors shops, Swapmeets, Fleamarkets and private collectors I should like to hear from any person who may have the odd, one-off item which may have bee produced, that I may have missed.

VEHICLES

MODEL	MB	MU	GC
No.M1541 Rolls Royce Silver Cloud.			
Metallic silver grey with silver wheels, radiator, grille, bumpers, lights and number plate, with off-white or golden interior. Price 19/6d. Issued 1960, Deleted 1969. 76mm.	£10	£7	£5
Also in light or medium sky blue, otherwise as above.	£8	£6	£4
No.M1542 Jaguar 3.4.			
Metallic medium or dark blue with full silver wheels, lights, radiator, bumpers etc. With dark blue interior. Price 19/6d. Issued 1960. Deleted 1969. 74mm.	£10	£7	£5
Also in dark red, but rare; otherwise as above.	£25	£20	£15
No.M1543 Humber Super Snipe.			
Metallic green with full silver wheels, lights, radiators, bumpers etc. Dark green interiors and like all the vehicles in this series it is reversible. Has registration plates and plate coach work accessories. Price 19/6d. Issued 1960. Deleted 1969. 74mm.	£10	£7	£5
Also in metallic bronze, otherwise as above.	£15	£10	£7

MODEL	MB	MU	GC

No.M1544 Luxury Coach.

Lime green, with darker green side panels, golden seats and interior
and full silver-plated wheels and trim etc. Price 25/-. Issued 1960.
Deleted 1969. 114mm.

	£25	£20	£15

Also in medium or dark yellow with darker orange or yellow
panels, otherwise as above.

	£18	£15	£10

No.M1545/R Double Decker Bus.

Red, with silver-plated wheels, grille, lights etc. With fawn or dark
brown seats and emblem in golden yellow 'London Transport' and
'Player's Taste Better' decals. Price 25/-. Issued 1961. Deleted 1970.
112mm.

	£35	£30	£25

Also in red with 'Go Well – Go Shell' decals in red and yellow,
otherwise as above.

	£20	£15	£10

No.M1545/G Double Decker Bus.

Green with dark green seats and interior. 'Green Line' emblems
and decals in dark gold or green and black and the number 715 in
black. Complete with full silver-plated wheels, grille, lights and
bumpers etc. Price 25/6d. Issued 1961. Deleted 1969. 112mm.

	£30	£25	£20

No.M1546 Lorry with Bale Load.

Red cab and wagon body with black chassis, full silver-plated
wheels, lights and bumpers. Complete with full bale load in blue.
Price 25/6d. Issued 1962. Deleted 1970. 120mm.

	£15	£12	£10

Also rose pink cab with dark blue chassis and matching bumpers,
otherwise as above.

	£20	£15	£10

No.M1547 Bedford Lorry with Container Load.

Rose pink cab and body with dark blue chassis and matching
bumpers. With blue container load and 'Global' decals in black on
orange background. Also the letter 'G' in orange circles. Price
25/6d. Issued 1962. Deleted 1970. 120mm.

	£18	£15	£12

Also dark blue cab and body, otherwise as above.

	£35	£30	£25

No.M1548 Car Transporter.

Red cab with medium blue chassis, mudguards and bumpers. Grey
and black transporter body and 'Car Transporters Ltd.' in black
with silver-plated wheels and lights. This colour is quite rare. Price
25/-. Issued 1967. Deleted 1970. 140mm.

	£40	£35	£30

MODEL	MB	MU	GC
With blue cab, black chassis, fawn carrier body and 'Haulport' decals, otherwise as above.	£30	£25	£20
With yellow cab and black or dark blue chassis, orange car transporter body and 'Europa–Transport'. Price 25/6d. Issued 1967. Deleted 1970. 140mm.	£20	£15	£10

No.M1549 Fire Chief's Humber.
Red, black and gold with 'Fire Chief' decals. Price 21/6d. Issued 1964. Deleted 1970. 74mm.	£15	£12	£9

No.M1550 Fire Engine with Emergency Light.
Red with silver ladder, grille, bumpers, wheels and warning bells on roof. Price 25/-. Issued 1963. Deleted 1970. 84mm.	£30	£25	£20
Red with silver ladder, 'Kent Fire Brigade' decals and white plastic wheels, otherwise as above.	£15	£12	£9

No.M1551 Shell Oil Tanker
Yellow cab and ladders, blue chassis and white tanker body, with 'Shell', 'Petroleum Products' and 'BP' decals and emblems in yellow, red, white and green. Price 25/-. Issued 1964. Deleted 1970. 120mm.	£35	£30	£25

No.M1552. Jaguar Police Car with Warning Light.
Dark blue with silver-plated wheels, grille, bumpers and lights. Police sign in black and white, roof light in blue. Price 21/6d. Issued 1963. Deleted 1970.	£15	£12	£10
Also in white with black and blue police decals and signs, otherwise as above.	£35	£30	£25

No.M1553. Caravan
Cream or dark yellow with off-white or cream interior. Price 2/11d. Issued 1964. Deleted 1967. 71mm.	£6	£5	£4
In green, otherwise as above.	£10	£7	£5
In blue. Issued 1967. Deleted 1970. Otherwise as above.	£9	£7	£4

No.M1554. Trailer for 1546 or 1547.
Golden yellow with silver-plated wheels and black tyres. Price 2/11d. Issued 1963. Deleted 1965. 81mm.	£5	£4	£3
Also in green, otherwise as above.	£7	£6	£5
Also in pink. Issued 1965. Deleted 1967. Otherwise as above.	£7	£6	£5

No.M1554/A. Flat Trailer.
Orange, silver and black. Price 4/-. Issued 1967. Deleted 1970. 56mm.	£5	£4	£3

No.M1555. Trailer With Boat.
White, blue and black with white and blue trailer and black tyres. Price 3/6d. Issued 1963. Deleted 1969. 69mm.	£7	£5	£3
Also in red, white and black, but rare. Otherwise as above.	£15	£10	£7

No.M1556. Mercedes Benz.
Golden yellow with matching interior, silver-plated wheels, grille, bumpers, lights etc. Price 19/6d. Issued 1963. Deleted 1969. 76mm.	£10	£7	£5
Also in metallic red, but rare. Otherwise as above.	£25	£20	£15

MODEL	MB	MU	GC

No.M1557. Austin A40.

Deep orange with black roof, and silver wheels, grille, bumpers, lights etc. Dark interior. Price 19/6d. Issued 1964. Deleted 1969. 71mm.

	£20	£15	£10

Also in black with orange roof, otherwise as above.

	£30	£25	£20

No.M1558. Mercedes Benz 300 SL.

White body with red roof and the number '52' in red on doors and bonnet, with silver-plated wheels, grille, bumpers, lights etc. Price 19/6d. Issued 1965. Deleted 1969. 76mm.

	£10	£7	£5

No.M1559. E-Type Jaguar.

Metallic green with large number '7' in black on white circle, on doors, although numbers may differ. Price 19/6d. Issued 1964. Deleted 1969. 76mm.

	£15	£22	£9

Also in metallic silver, otherwise as above.

	£25	£20	£15

No.M1560. Renault Floride.

Yellow with black roof, and silver-plated wheels, bumpers and lights. Price 19/6d. Issued 1964. Deleted 1969. 74mm.

	£10	£7	£5

In bright orange with black roof, otherwise as above.

	£20	£15	£10

No.M1561. Morris 1000.

Red and black, with full silver trim. Price 16/11d. Issued 1965. Deleted 1967. 74mm.

	£30	£25	£20

No.M1562. Mobile Bank.

Blue with gold lettering 'Security Is Best'. Rare model. With full silver trim, opening rear doors, to reveal sacks, bars of gold and other coinage. Price 21/-. Issued 1965. Deleted 1967. 91mm.

	£50	£45	£35

No.M1563. Securicor Van.

Black or blue with authentic 'Securicor' decals and emblems. Price 21/-. Issued 1965. Deleted 1967. 91mm.

	£50	£45	£35

No.M1564. Steam Lorry.

Medium or light blue, this old-time special had the emblems and decals, 'Coal Ayers Limited'. With silver-plated wheels and lights and golden funnel, dark cab interior. Price 21/-. Issued 1964. Deleted 1969. 91mm.

	£30	£25	£20

Also in green, otherwise as above.

	£35	£30	£25

No.M1565. Breakdown Lorry With Warning Light.

Blue or purple cab with yellow roof light, yellow and red body with blue crane and trailer wheels. White plastic wheels and bumpers, with the 'Day and Night Service — Jackson Motors' decals in black on red background. Price 25/-. Issued 1963. Deleted 1969. 76mm. With trailer 99mm.

	£15	£12	£9

No.M1565/A. Breakdown Lorry.

Silver cab with orange body or deep yellow body with orange crane and 'Minic Motor Co.' decals and BP emblem in red and green. Price 17/6d. Issued 1966. Deleted 1969. 76mm.

	£15	£12	£9

No.M1566. Delivery Van.

Red, blue or black with various names and decals. Popular one being 'Road Services Maintenance'. Price 17/6d. Issued 1965. Deleted 1967. 74mm.

	£25	£20	£15

No.M1567. 3.4 Jaguar.

Metallic green British racing colours and the number '10' in black on white circle. Price 19/6d. Issued 1963. Deleted 1969.

	£14	£12	£10

No.M1568. 3.4 Jaguar. Continental.

Brilliant white or dark grey with the number '10' in red on white circle, these being continental colours. Price 19/6d. Issued 1963. Deleted 1967. 74mm.

	£20	£15	£10

No.M1569. Conqueror Tank.

Military green with 'Army' decals and chevrons, and swivelling gun tower with thick tracks. Price 19/6d. Issued 1962. Deleted 1967. 91mm.

	£20	£15	£10

No.M1570. Mechanical Horse.

Red, yellow and blue, adaptable for fixing to trailers. Made especially for work in station goods yards. Price 10/6d. Issued 1965. Deleted 1969. 67mm approx.

	£9	£7	£5

No.M1571. Station Trolley.

Yellow and black, made for carrying parcels, mail and other heavy items. Price 10/6d. Issued 1965. Deleted 1969. 68mm.

	£20	£15	£10

No.M1572. Royal Mail Van.

Red with gold lettering and emblem. Only a limited number were made. Price 17/6d. Issued 1965. Deleted 1967. 71mm.

	£50	£45	£35

No.M1573. Aston Martin DB4.

Aston Martin racing livery as it appeared when it won the 1959 World Constructors Championship. This model was an immediate success. With Italian body designed by Superleggera Touring and built under licence in Great Britain. Also had various numbers including the '8' attached to it. Price 16/1d. Issued 1966. Deleted 1969.

	£20	£15	£10

No.M1574. Porsche Carrera GT.

This famous racing car was in white or cream with the number '4' in white on a black circle. Full silver-plated wheels, bumpers, lights etc. Designed by Dr. Ferdinand Porsche, who had based this car on his earlier Volkswagen design. The success of this car was highly noted, especially by winning the French Grand Prix 1962, the Targa Florio 1964 and gaining second place in the 1965 Targa. Price 16/11d. Issued 1964. Deleted 1969. 60mm.

	£15	£10	£7

No.M1575. Daimler Special.

Red, black and gold with spoked wheels and the number '21' in black in white circles. Made for the International Toy Year and available in a limited edition in a special presentation box. Price 9/6d. Issued 1966. Deleted 1969. 76mm.

£50   £45   £35

No.M1576. Ferrari 500 Superfast.

Of all the world's sports car, unequalled in its heyday, for mystique and glamour. It was the dream car that only a chosen few could afford. Enzo Ferrari's customers were the world's rich and famous. In brilliant rose pink or the rare deep maroon which is worth double. Spoked wheels and the number '21' in black on white circles. For those interested, the price of the actual racing car was £10,932. Price 17/6d. Issued 1965. Deleted 1969. 76mm.

£20   £15   £10

No.M1577. Chevrolet Corvette Stingray.

Golden yellow with off-white or cream interior, black grille, and bumpers and white plastic wheels. When the Corvette began in 1953, it was the world's first car in relatively large scale production to have a body of glass-reinforced polyester. It had a tuned 8-cylinder Chevrolet engine of just under 4 litres. Price 19/6d. Issued 1966. Deleted 1969. 74mm.

£10   £7   £5

Also in white with silver-plated wheels, bumpers, lights and black grille, otherwise as above.

£15   £10   £7

No.M1578. Car Water Ferry.

Dark brown and grey with red and white barrier gates. Any Minic Motorway car can drive on or off. Price 7/6d. Issued 1964. Deleted 1969.

£7   £5   £3

No.M1579. Car Water Loading Quay.

A fine edition to the Motorway range, lime green, darker green and grey. Red and white barrier gate, steps and chains. Price 17/6d. Issued 1964. Deleted 1969.

£10   £7   £5

No.M1580. Car Water Ferry.

Dark brown, lime green, dark green and grey with red and white barrier gates in a special presentation box. Price 27/6d. Issued 1965. Deleted 1969.

£15   £10   £7

No.M1581. Aston Martin DB6.

Metallic green with dark interior, metal wheels, bumpers, grille, lights etc. Price 22/6d. Issued 1968. Deleted 1970. 74mm.

£10   £7   £5

No.M1582. E-Type Jaguar. 2 x 2.

Rose pink body with off-white or black interior, silver bumpers and wheels and black grille, streamlined headlights. Price 22/6d. Issued 1969. Deleted 1970. 74mm.   £10   £7   £5

No.M1587. Ford GT MK 11.

Metallic orange or deep gold, which is worth double, with black grille, and silver wheels. This car was the 1966 winner of Le Mans ' 24 hours', Sebring '12 hours' and Daytona '500'. Price 22/6d. Issued 1968. Deleted 1970. 74mm.   £10   £7   £5

No.M1588. Ferrari 330-P4.

Rose pink with black grille, silver wheels and dark interior. This car was the world's sports car champion winner in 1967. A very successful model. Price 24/-. Issued 1968. Deleted 1970. 76mm.   £12   £9   £7

No.M1589. Alfa Romeo T33.

Deep red or the rare purple colour which is worth treble. With silver motive on bonnet and silver wheels. This car was the winner of the Mugello Endurance Race in 1968 and gained second place in the Targa Florio the same year. Price 23/6d. Issued 1969. Deleted 1970. 76mm.   £10   £7   £5

No.M1590. Porsche 907.

Brilliant white with silver wheels. The 1968 winner of the Daytona 500, the Sebring '12 hours', the Targa Florio and the Nuremberg 100 KMS. Price 25/-. Issued 1969. Deleted 1970. 76mm.   £15   £10   £7

CARPLAY SERIES

MODEL	MB	MU	GC
M/166. Super Track 4 Plans Booklet. Red, yellow and blue with 'Triang' decals. Shows how to build up your layout for a basic train set. Has 36 pages of ideas, plans, track geometry, dimensions, prices and wiring diagrams. Invaluable for the beginner and valuable for the expert. Price 2/-. Issued 1965. Deleted 1970.	£3	£2	£1
M/395. Triang Hornby Sounds Record. In multicoloured jacket, this is a 7-inch extended play 45 rpm record. Price 3/11d. Issued 1964. Deleted 1970.	£3	£2	£1
M/1700. Road Sign Base. White or silver. Price 2d. Issued 1963. Deleted 1970.	25p	15p	10p
M/1710. Telegraph Pole. Black or grey blue for set of six. Price 1/9d. Issued 1963. Deleted 1970.	50p	30p	20p
M/1711. Street Light, Single. Cream and orange with black base. Price 9d. Issued 1963. Deleted 1970.	40p	30p	20p
M/1712. Street Light, Double. Cream and orange with black base. Price 1/-. Issued 1963. Deleted 1970.	60p	50p	30p
M/1715. R.A.C. Patrol Box. Blue with 'R.A.C.' decals. Price 1/3d. Issued 1963. Deleted 1970.	£1	75p	50p
M/1716. A.A. Patrol Box. Grey, white and black with 'A.A.' decals. Price 1/6d. Issued 1964. Deleted 1970.	£1	75p	50p
M/1717. P.O. Telephone Booth. Red. Price 1/6d. Issued 1963. Deleted 1970.	£1	75p	50p
M/1718. Police Telephone Booth. Blue with 'Police' decals. Price 1/6d. Issued 1963. Deleted 1970.	£1	75p	50p
M/1719. P.O. Pillar Box. Red and black with 'P.O.' decals. Price 6d. Issued 1963. Deleted 1970.	50p	30p	20p
M/1720. Bus Passenger Shelter. Cream or fawn with 'Bus Stop' sign in red and black. Price 2/-. Issued 1963. Deleted 1970.	£1	75p	50p
M/1721. Set of 6 Road Signs. Green, red and black, complete with base clips. Price 2/6d. Issued 1964. Deleted 1970.	£2	£1.50	£1
M/1725. Traffic Lights. Black, white, amber, green and red. Price 1/3d. Issued 1964. Deleted 1970.	£1	75p	50p
M/1733. Set of Lamp Standards. Green, orange and black. Contains four single and two double light standards with clip on bases. Price 5/-. Issued 1964. Deleted 1970.	£1.50	£1	50p

MODEL	MB	MU	GC
M/1734. P.O. Police and Telegraph Set.			
Black, red and blue. Contains police box, P.O. telephone box and six telegraph poles. Price 5/6d. Issued 1964. Deleted 1970.	£3	£2	£1
M/1735. R.A.C./A.A. Set.			
Black, blue, yellow and white. Contains R.A.C. box, A.A. box and six telegraph poles with clip-on bases. Price 5/-. Issued 1964. Deleted 1970.	£3	£2	£1
M/1746. Speed Controller.			
Red and black, complete with leads. Price 7/6d. Issued 1964. Deleted 1970.	£2.50	£2	£1
M/1747. Battery Container and Contacts.			
Blue with 'Triang' decals and wording in yellow and white. Price 1/-. Issued 1964. Deleted 1970.	50p	30p	20p
M/1811. Heliport.			
Very light blue, dark blue, black and white, complete with working helicopter which is launched by a car passing through the building. Price 12/6d. Issued 1966. Deleted 1970.	£7	£5	£3
M/2001. Filling Station.			
White with pinkish tint with 'Shell' decals and signs. Complete with black and orange roadway track. Price 10/2d. Issued 1963. Deleted 1970. 228mm x 76mm.	£9	£7	£5
M/2002. Service Station.			
Rose pink, white and black with 'Shell' decals. Price 13/9d. Issued 1963. Deleted 1970. 228mm x 168mm.	£9	£7	£5
M/2003. Drive-Through Building.			
Rose pink, white and black with fawn base. Price 18/-. Issued 1963. Deleted 1970. 228mm x 168mm.	£10	£8	£6
M/2004. Bungalow with Garage.			
White, black and orange with blue door. Price 16/3d. Issued 1963. Deleted 1970. 228mm x 168mm.	£10	£8	£6
M/2005. Double Motel Chalet With Garages.			
Orange, white, black, green and grey livery. Price 13/9d. Issued 1964. Deleted 1970. 228mm x 168mm.	£14	£12	£10
M/2006. Transport Depot.			
Pink, lime green, white, black and fawn. Price 15/-. Issued 1964. Deleted 1970. 228mm x 168mm.	£9	£7	£5
M/2007. Racing Pit.			
White, rose pink, black and blue. Complete with a variety of advert decals, including 'Porsche', 'Mercedes Benz', 'Jaguar', 'K.L.G.' and 'Dunlop'. Price 9/1d. Issued 1964. Deleted 1970. 152mm x 89mm.	£5	£4	£3
M/2008. Freight Depot.			
Orange, blue, black and yellow with Freight Depot decals. Price 11/9d. Issued 1964. Deleted 1970. 228mm x 168mm.	£10	£8	£6

TRACK

MODEL	MB	MU	GC
RM901. Railway/Motorway Track Level Crossing.			
Black and orange, ungated. Price 8/11d. Issued 1963. Deleted 1970. 15.2cms approx.	£3	£2	£1
RM903. Two Gates and Pillars.			
White and red. Made for RM901 level crossing in 1963. Price 2/6d. Issued 1963. Deleted 1970. 15.2cms.	£2	£1.50	£1
RM904. Set of Two Lifting Booms and Pillars.			
Black, orange and white. Made for RM901. Price 2/6d. Issued 1963. Deleted 1970.	£1	75p	50
RM905. Double Track Level Crossing.			
Black, orange, white and green. Price 12/6d. Issued 1964. Deleted 1970. Approx. 26cms.	£5	£4	£3
RM910. Straight Track.			
Price 4/3d. Issued 1963. Deleted 1970. 15.2cms.	75p	50p	35p
RM911. Road/Rail Junction Type A.			
Orange and black. Price 7/-. Issued 1963. Deleted 1970. Radius 15.2cms.	£3	£2	£1
RM912. Road/Rail Junction Type B.			
Orange and black. Price 7/-. Issued 1963. Deleted 1970. Radius 15.2cms.	£3	£2	£1
RM913. Road/Rail Buffer Stop (Left).			
Black, red and brown. Price 2/11d. Issued 1963. Deleted 1970. 7.6cms.	£2	£1.50	£1
RM914. Road/Rail Buffer Stop (Right).			
Black, red and brown. Price 2/11d. Issued 1963. Deleted 1970. 7.6cms.	£2	£1.50	£1
RM921. Car Loading Ramp.			
Black, orange and green. Price 15/6d. Issued 1963. Deleted 1970. 32cms approx.	£7	£5	£3
RM922. Railway Car Transporter.			
Orange, grey and black. Price 13/6d. Issued 1963. Deleted 1970. 30cm approx.	£3	£2	£1
RM923. Railway Adaptor Bogie.			
Black and silver/grey. Price 3/11d. Issued 1963. Deleted 1970. 15.2cms.	£1	£1.50	£1
RM924. Road/Rail Wagon.			
Brown, black and gold with British Railway decals. Price 2/11d. Issued 1963. Deleted 1970. 7.5cms.	£3	£2	£1
M1601. Straight Standard.			
Orange and black. Price 2/6d. Issued 1963. Deleted 1970. 15.2cms.	£1.50	£1	50p
M1602. Half Straight.			
Orange and black. Price 1/11d. Issued 1963. Deleted 1970. 7.6cms.	£1	75p	50p

MODEL	MB	MU	G(
M1603. Straight Starting Grid. Orange and black. Price 2/6d. Issued 1963. Deleted 1970. 15.2cms.	£2	£1.50	£1
M1604. Straight Power Pick-Up. Orange and black. Price 4/11d. Issued 1963. Deleted 1970. 15.2cms.	£2	£1.50	£1
M1605. Zebra Crossing with Beacons. Orange and black with non-flashing beacons and two pavement sections in fawn and green. Price 6/-. Issued 1963. Deleted 1970. 15.2cms.	£4	£3	£2
M1606. Straight Single Track. Orange and black. Price 1/9d. Issued 1963. Deleted 1970. 15.2cms.	£1.50	£1	**50**
M1607. Straight Single Track Power Pick-Up. Orange and black. Price 2/11d. Issued 1963. Deleted 1970. 15.2cms.	£1.50	£1	50p
M1608. Straight Hill Grip, Standard. Orange and black. Price 2/11d. Issued 1963. Deleted 1969. 15.2cms.	£1.50	£1	50p
M1609. Double Straight Standard. Orange and black. Price 4/6d. Issued 1963. Deleted 1970. 30.4cms.	£3	£2	£1
M1610. Slate Chicane Standard. Orange and black. Price 3/11d. Issued 1963. Deleted 1970. 15.2cms.	£1.50	£1	50p
M1611. Bend (Large). Orange and black. Price 3/11d. Issued 1963. Deleted 1970. Radius 15.2cms.	£1.50	£1	50p
M1613. Bend (Small). Orange and black. Price 2/11d. Issued 1963. Deleted 1970. Radius 7.5cms.	£1	75p	50p
M1615. Bend (Very Small). Orange and black. Price 2/6d. Issued 1963. Deleted 1970. Radius 3.6cms.	50p	40p	30p
M1616. Outer Bend. Orange and black. Price 3/6d. Issued 1963. Deleted 1970. Radius 25cms.	£1.50	£1	50p
M1617. Single Track Bend. Orange and black. Price 2/6d. Issued 1963. Deleted 1970. Radius 7.8cms.	£1	75p	50p
M1618. Straight Track/One Side Isolating. Orange and black. Price 2/9d. Issued 1963. Deleted 1970. 15.2cms.	£1	75p	50p
M1619. Straight With One Side Actuating. Orange and black. Price 2/9d. Issued 1963. Deleted 1970. 15.2cms.	£1	75p	50p
M1620. Relay With Leads. Brown. Price 4/11d. Issued 1963. Deleted 1970. 7.6cms.	£2	£1.50	£1

MODEL	MB	MU	GC
1621. Roundabout. Black and orange. Price 29/9d. Issued 1963. Deleted 1970.	£7	£5	£3
1623. Crossroads. Black and orange. Price 6/6d. Issued 1963. Deleted 1970. 30.4cms.	£3	£2	£1
1624. Right Junction. Black and orange. Price 9/11d. Issued 1963. Deleted 1970. 22.9cms. Radius 15.2cms.	£3	£2	£1
1626. Left Junction. Black and orange. Price 9/11d. Issued 1963. Deleted 1970. Length 22.9cms. Radius 15.2cms.	£3	£2	£1
1627. Humpedback Bridge Summit. Black and orange. Price 3/6d. Issued 1963. Deleted 1970. 15.2cms.	£3	£2	£1
1628. Gradient Base. Black and orange. Price 3/6d. Issued 1963. Deleted 1970. 15.2cms.	£3	£2	£1
1629. Gradient Summit. Black and orange. Price 3/6d. Issued 1963. Deleted 1970. 15.2cms.	£3	£2	£1
1630. Y Junction. Orange and black. Price 8/9d. Issued 1963. Deleted 1970. Radius 15.2cms.	£4	£3	£2
1631. Straight Change-Over Track. Black and orange. Price 3/6d. Issued 1963. Deleted 1970. 15.2cms.	£3	£2	£1
1632. Junction, Right, Lay-By. Orange and black. Price 7/11d. Issued 1963. Deleted 1970. 15.2cms.	£3	£2	£1
1633. Junction, Left, Lay By. Orange and black. Price 7/11d. Issued 1963. Deleted 1970. 15.2cms.	£3	£2	£1
1634. Straight Single Track Starting Grid. Orange and black. Price 1/11d. Issued 1963. Deleted 1970. 15.2cms.	£1.50	£1	50p
1635. Gradient Base. Single Track. Orange and black. Price 2/6d. Issued 1963. Deleted 1970. 15.2cms.	£1.50	£1	50p
1636. Gradient Summit. Single Track. Orange and black. Price 2/6d. Issued 1963. Deleted 1970. 15.2cms.	£1.50	£1	50p
1637. Half Straight Single Track. Orange and black. Price 1/6d. Issued 1963. Deleted 1970. 7.6cms.	75p	50p	25p
1638. Quarter Straight Standard. Orange and black. Price 1/6d. Issued 1963. Deleted 1970. 3.8cms.	50p	30p	20p
1639. Zebra Crossing with Flashing Beacon and Two Pavement Sections. Orange, black and white, fawn and green. Price 8/6d. Issued 1964. Deleted 1970. 15.2cms.	£7	£5	£3

MODEL	MB	MU	G(
M1640. Straight Pavement Section Without Steps. Fawn and green. Price 1/-. Issued 1963. Deleted 1970. 15.2cms.	£1	75p	50,
M1641. Straight Pavement Section With Steps. Fawn and green. Price 1/-. Issued 1963. Deleted 1970. 15.2cms.	£2	£1.50	£1
M1642. Half Straight Pavement Section. Fawn and green. Price 9d. Issued 1963. Deleted 1970. 7.6cms.	50p	30p	20,
M1643. Pavement Standard Bend. (Inner). Fawn and green. Price 1/-. Issued 1963. Deleted 1970. Radius 15.2cms.	£1	75p	50,
M1644. Pavement Standard Bend. (Outer). Fawn and green. Price 1/-. Issued 1963. Deleted 1970. Radius 15.2cms.	£1	75p	50,
M1645. Right Pavement Junction. Fawn and green. Price 1/-. Issued 1963. Deleted 1970. 15.2cms.	£1	75p	50,
M1646. Left Pavement Junction. Fawn and green. Price 1/-. Issued 1963. Deleted 1970. 15.2cms.	£1	75p	50,
M1647. Pavement Crossroads. Fawn and green. Price 1/-. Issued 1963. Deleted 1970. 15.2cms.	£1	75p	50,
M1648. Pavement Roundabout. Fawn and green. Price 1/-. Issued 1963. Deleted 1970. 15.2cms.	£1	75p	50,
M1649. Pavement Bend, Standard Inner. Fawn and green. Price 1/-. Issued 1963. Deleted 1970. Radius 7.5cms.	75p	50p	25p
M1650. Pavement Bend, Standard Outer. Fawn and green. Price 1/-. Issued 1963. Deleted 1970. Radius 7.5cms.	75p	50p	25p
M1651. Pavement Bend, Standard Inner. Fawn and green. Price 1/-. Issued 1963. Deleted 1970. Radius 7.5cms.	75p	50p	25p
M1653. Left Pavement End. Fawn and green. Price 1/-. Issued 1963. Deleted 1970. 7.5cms.	£1	75p	50p
M1654. Right Pavement End. Fawn and green. Price 1/-. Issued 1963. Deleted 1970. 7.5cms.	£1	75p	50p
M1655. Pavement Gradient Base 'Left'. Green and fawn. Price 1/-. Issued 1963. Deleted 1970. 15.2cms.	75p	50p	25p
M1656. Pavement Gradient Base 'Right'. Green and fawn. Price 1/-. Issued 1963. Deleted 1970. 15.2cms.	75p	50p	25p
M1657. Pavement Gradient Summit, 'Left & Right'. Green and fawn. Price 1/-. Issued 1963. Deleted 1970. 15.2cms.	£1	60p	40p
M1658. Pavement Bridge & High Levels Straight. Green and fawn. Price 1/-. Issued 1963. Deleted 1970. 15.2cms.	£1	60p	40p

MODEL	MB	MU	GC
M1659. Pavement Junction 'Y' Left.			
Green and fawn. Price 1/-. Issued 1963. Deleted 1970. 15.2cms.	£1	60p	40p
M1660. Pavement Junction 'Y' Right.			
Green and fawn. Price 1/-. Issued 1963. Deleted 1970. 15.2cms.	£1	60p	40p
M1661. Pavement Bend Outer.			
Green and fawn. Price 1/-. Issued 1963. Deleted 1970. 15.2cms.	£1	60p	40p
M1662. Pavement Bend Inner.			
Green and fawn. Price 1/-. Issued 1963. Deleted 1970. 15.2cms.	£2	£1.50	£1
M1663. Pavement High Level Bend, Outer.			
Green and fawn. Price 1/-. Issued 1963. Deleted 1970. 15.2cms.	£2	£1.50	£1
M1664. Pavement High Level Bend, Inner.			
Green and fawn. Price 1/-. Issued 1963. Deleted 1970. 7.8cms.	£2	£1.50	£1
M1665. Pavement for Humpedback Bridge Summit.			
Green and fawn. Price 1/-. Issued 1963. Deleted 1970. 15.2cms.	£2	£1.50	£1
M1670. Flyover Set.			
Contains M1601 straight, 4 M1608 straight hill grips; 2 M1628 gradient bases; 2 M1629 gradient summits; 8 M1675 gradient pier caps; 2 M1676 gradient piers No.2; 2 M1677 gradient piers No. 3 and 2 M1678 gradient piers No.4. Price 35/-. Issued 1963. Deleted 1970.	£15	£12	£10
M1671. Gradient Pier Set.			
Contains 4 M1675 gradient pier caps, 1 M1676 gradient piers; No.2, 1 M1677 gradient pier No. 3, and 1 M1678 gradient pier No. 4. Price 4/6d. Issued 1963. Deleted 1970.	£2	£1.50	£1
M1672. Humpedback Bridge Set.			
Black, white and green. Contains 2 M1608 straight hill grips, 1 M1627 humpbacked bridge summit, 2 M1628 gradient bases, 4 M1675 gradient pier caps, and two M1676 gradient piers No.2. Price 18/5d. Issued 1963. Deleted 1970.	£10	£7	£5
M1673. Pier Set for Humpedback Bridge.			
Contains 4 M1675 gradient pier caps and 2 M1576 gradient piers No.2. Price 3/6d. Issued 1963. Deleted 1970.	£1.50	£1	50p
M1675. Gradient Pier Cap.			
Green. Priced 1d. Issued 1963. Deleted 1970.	30p	20p	10p
M1676. Gradient Pier No. 2.			
Green. Price 2d. Issued 1963. Deleted 1970.	30p	20p	10p
M1677. Gradient Pier No. 3.			
Green. Price 3d. Issued 1963. Deleted 1970.	40p	30p	20p
M1678. Gradient Pier No. 4.			
Fawn, lime green or cream. Price 4/6d. Issued 1963. Deleted 1970.	£1	75p	50p
M1678/A. Gradient Pier No. 4 with Cap Set.			
Fawn, lime green or cream. Price 8/9d. Issued 1963. Deleted 1970.	£3	£2	£1

MODEL	MB	MU	G
M1682. Bridge Pylon With Pier Car.			
Lime green or fawn. Price 1/6d. Issued 1963. Deleted 1970.	£1	75p	50
M1683. Bridge Pylon Height Extension Set.			
Lime green or fawn. Price 3/-. Issued 1963. Deleted 1970.	£1.50	£1	50
M1690. Grass Strip Centre Straight.			
Green and fawn. Price 1/-. Issued 1963. Deleted 1970. 15.2cms.	£3	£2	£1
M1691. Grass Strip Centre Bend.			
Green and fawn. Price 1/-. Issued 1963. Deleted 1970. 15.2cms.	£3	£2	£1
M1692. Grass Strip Centre End.			
Green and fawn. Price 1/-. Issued 1963. Deleted 1970. 3.8cms.	50p	40p	25
M1693. Grass Centre Roundabout.			
Green and fawn. Price 1/-. Issued 1963. Deleted 1970. Radius 9cms.	£2	£1.50	£1

MINIC MOTORWAY SETS

MODEL	MB	MU	GC

M925. Road/Rail Set.
One of the best investments in this section, consists of M1570 mechanical horse in yellow and maroon, a railway adaptor bogie RM923 in grey and one road and rail wagon in matching maroon with British Railways decals. Price 33/2d. Issued 1964. Deleted 1970.   £40

M1504. Saloon Car Set.
Contains lime green Rolls Royce Silver Cloud, black Humber Super Snipe, three straight rails, a power pick up straight, four bends and two speed and direction controllers. Price 85/11d. Issued 1964. Deleted 1970. 81 x 51cms approx. space layout.   £25

M1506. Great Championship Set.
Contains pink Ford racing car and orange Ferrari, although colours may vary. Two hand controllers and two track mates (with 305 cms of track which makes up into an up-and-over figure of eight layout with a large fast loop and a tight bend to test out skilful drivers). Complete with track support piers. Price 95/-. Issued 1969. Deleted 1970.   £20

M1507. Criss-Cross Racing Set.
Contains a Ford and a Ferrari racing car, colours may vary, plus two hand controllers and two tack mates with over 150cms. of track in an oval. Two cross-over tracks give this set added excitement. Price 89/11d. Issued 1969. Deleted 1970.   £15

M1508. Fire-fighters Set.
This rare set contains a fire chief's car and a fire engine in matching red, two hand controllers, two track mates with more than 305cms. of track. plus a large fire station and a set of firemen. Complete with track support piers. Price 79/11d. Issued 1966. Deleted 1969.   £50

M1509. The Travellers Set.
Contains one cream Mercedes Benz and matching cream caravan, two hand controllers, almost 210cms. of track, and with various layouts, packets of scenery, shrubs etc. Price 45/-. Issued 1965. Deleted 1970.   £20

M1510. Rough Riders Rally Set.
Contains a Ferrari 500 and a Chevrolet Stingray in white and pink, although colours may vary. Two hand controllers, two track mates, a mad motorist hazard and almost 300cms. of track, including chicane and bridge supports. Price 157/6d. Issued 1968. Deleted 1970.   £25

M1511. Commercial Trunker Set.
Contains a Bedford Lorry in yellow or lime green with black chassis and blue container load and a Shell Oil Tanker in matching yellow, lime green and blue with 'Petroleum Products', 'Shell' and 'B.P.' decals. Three straights, four bends, one straight power pick-up and two speed and direction controllers. Price 94/6d. Issued 1964. Deleted 1970. 81 x 57cms layout space required.   £35

M1512. Public Transport Set.
Contains a red double decker bus with 'Players Taste Better' advert on sides and one blue and brown coach, although colours and adverts may vary. Three straights, four half straights, six bends,

one cross roads, one straight power pick-up and two speed and
direction controllers. Price 113/6d. Issued 1964. Deleted 1970.
112cm x 86cms layout space required.   £40

M1513. International Golden 500 Set.

Contains a Ferrari 500 and a Chevrolet Stingray in pink and
orange, though colours may vary, two hand controllers, an
automatic starting gate, two chicanes, two track mates and over
11 ft. (330cms.) of track. To start the race, press a button at the
back of the starting gate. This lowers the Union Jack and
automatically switches on the power. In the event of a crash, press
another button which lowers the red flag and switches off the
power. A third button operates the chequered flag to signal the
winner. Price 159/6d. Issued 1967. Deleted 1970.   £30

M1514. European Silver 8 Set.

Contains an Alfa Romeo and a Porsche, one pink and one white,
although colours may vary. Two hand controllers and almost
210cms. of track, including chicane to build an up-and-over figure
eight or an exact replica of the Grand Piano circuit. Price 159/6d.
Issued 1969. Deleted 1970.   £25

M1515. Frontier Post Rally Set.

An exciting set with two car big track. Complete with a Frontier
Post that really works. The barrier goes up and down across the
track automatically and the cars race to get through before the
frontier closes. Complete with a Mercedes 220 in yellow and a 3.8
Jaguar in pink or red, plus a sensational track which makes into no
less than seven different layouts. Price 182/-. Issued 1967. Deleted
1970.   £25

M1516. Check Point Bravo Set.

Contains white Mercedes and red Jaguar, two hand controllers, two
track mates, a Rally check-point with barrier and filling station.
Also contains 270cms. of track, including chicane. Colours may
vary. Price 134/-. Issued 1969. Deleted 1970.   £20

M1517. Frontier Post Set No. 2.

Contains a yellow, green and grey customs barrier and a clockwork
motor which raises and lowers the barrier. Complete with two rally
cars, one pink and one white. A various amount of track which
makes no less than five different layouts, plus a neat little service
station with 'Shell' decals. Price 103/6d. Issued 1969. Deleted 1970.   £30

M1518. Holiday Highway Set.

Contains a green Humber saloon and a bright orange caravan, a
Bedford lorry with container load in red, black and blue, two hand
controllers and over 6 ft. (150cms.) of roadway to form an oval and
bottle neck. Price 103/6d. Issued 1968. Deleted 1970. £35

M1519. Crime Patrol Set.

Contains a Police Jaguar in white with roof light and a Jaguar in
red, two hand controllers, two track mates, a police station and
nearly 210cms. of roadway with bottle-neck and junction. Price
103/11d. Issued 1968. Deleted 1970. £30

M1520. Securicor Set.

Contains a blue Securicor van, a Police Jaguar with roof light, two
hand controllers, two track mates, a bank and small sacks of gold.
Nearly 7 ft (210cms.) of roadway. Price 123/11d. Issued 1968.
Deleted 1970. £50

M1522. Racing Set.

Contains a Mercedes 300 SL in white or silver with red roof and
the number 52 on doors and bonnet. Also a green E-type Jaguar,
four straights, two half straights, a straight starting grid, a straight
power pickup, six bends, a humpbackeded bridge summit, two
gradient bases, four gradient pier caps, two gradient piers No.2.
and two speed and direction controllers. Price 113/6d. Issued 1964.
Deleted 1970. 127 x 66cms layout space required. £25

M1525. Economy Racing Set.

Contains an Aston Martin DB4 and a Porsche Carrera in cream
and red, although colours may vary. One straight starting grid, a
straight power pick-up, eight bends, four gradients, two wire-
formed piers, six track-fitting clips, a set of fencing, bases for
fencing and two speed controllers. Price 79/11d. Issued 1964.
Deleted 1970. 81 x 66cms layout space required £25

M1526. Three Lane Racing Set.

Contains a Mercedes Benz, an E-type Jaguar and a Jaguar 3.4 in
green, red and silver with red roof, although colours may vary. A
straight starting grid, a straight power pick-up, single track, four
bends, eight outer bends, sixteen bends (single track), straight
starting grid, (single track), four double pier caps, four single pier
caps, three pairs of legs, left and right, and three speed and
direction controllers. Various layouts can be made. Price 182/-.
Issued 1964. Deleted 1970. 122 x 96cms approx. £40

M1527. Heliport International Set.

Contains Aston Martin in red, a break-down lorry in blue, orange
and red, a heliport building in blue, white, black and yellow,
complete with helicopter in pink, blue and black. Two hand
controllers, two track mates and over 150cms. of roadway to make
an oval. Wind spool on roof of heliport and locate helicopter into
cradle. As vehicle passes beneath it it automatically releases
helicopter which soars upwards and really flies! Price 130/-. Issued
1967. Deleted 1970. £35

M1528. Cobra Racing Set.

Contains two racing cars (colours may vary), two hand controllers,
two track mates and a mad motorist hazard, complete with garage.
The mad motorist in the first racing non-electric car reverses out of
his garage across the road (by means of a long running clockwork

mechanism), sees the oncoming cars and drives back again. You must decide whether to accelerate past him or brake. Price 189/-. Issued 1968. Deleted 1970.   £30

M1530. Trophy Accessory Set.

Contains Dunlop footbridge in grey and white, a pit building in green and white, flags, fences and start and finish banners. Price 17/6d. Issued 1968. Deleted 1970.   £5

M1531. International Circuit Extension Set.

Contains an automatic starting gate, a pit building in green and white, flags, fences and 61cms. of track, including chicane. Price 33/9d. Issued 1969. Deleted 1970.   £10

M1532. Cobra Track Layout Extension Set.

Contains a very large section of track giving a running length of 610cms. all within the amazingly small space of 150cms. x 105cms. Complete with all accessories and one of the best tracks in the whole series. Price 119/6d. Issued 1969. Deleted 1970.   £15

M1622. Automatic Cross Roads Set.

Contains one cross roads, two straights with one side isolated, four straights with one side actuating and one relay and set of connecting wires. Price 39/6d. Issued 1964. Deleted 1970.   £7

M1668. Flyover Set.

Contains all the components to build a flyover approximately 135cms. long. Essential for the real enthusiast. Price 50/-. Issued 1969. Deleted 1970.   £10

M1669. Humpbackeded Bridge Set.

Contains all the components to make a bridge approximately 75cms. long. Price 63/-. Issued 1969. Deleted 1970.   £15

M1706. Set of Flags.

Contains six flags of various nations. Price 1/11d. Issued 1964. Deleted 1970.   £1

M1726. Set of Warning Signs.

Contains cross roads, roundabout, bend, T-junction, Z-bend and sign. Price 2/8d. Issued 1964. Deleted 1970.   £1

M1727. Set of Instruction Signs.

Six more road signs, in authentic colours, which include a 'No Waiting', 'Bus Stop', 'No Entry', Inward and Outward, a 30-mile limit and a 40-mile limit. Price 2/8d. Issued 1964. Deleted 1970.   £1

M1728. Six More Instruction Signs.

Contains two 'Halt' signs, two 'Keep Left' signs and two 'Slow Major Road Ahead' signs. Price 2/8d. Issued 1964. Deleted 1970.   £1

M1729. Set of Motorway Signs.

Contains two very large Motorway signs, two medium Motorway signs and two small Motorway signs set out on a red card and contain two which vary according to districts. Price 2/8d. Issued 1964. Deleted 1970.   £2

M1730. Set of Town and Country Direction Signs.

Contains two 'Lay By' signs, one 'Leeds and York' sign, one

'London and Reading' sign and two other country signs which vary. Price 2/8d. Issued 1964. Deleted 1970.　　　£2

M1736. Flags and Fences Set.

Contains set of fencing and six flags of various nationalities, complete with clip-on bases. Price 15/-. Issued 1969. Deleted 1970.　　　£5

M1748. Speed and Direction Control Set.

Contains one black and one red speed controller, complete with good length of wiring in special presentation box. Price 9/11d. Issued 1964. Deleted 1970.　　　£5

M1749. Speed and Direction Control Set No.2.

Contains two black speed controllers and one direction controller in red, complete with leads. Price 17/6d. Issued 1964. Deleted 1970.　　　£10

M1757. Service Tool Kit Set.

Contains tools and a comprehensive range of spare parts to ensure that rally, racing and motorway vehicles will always be ready to go on the road. In special blue, white and red box. Price 10/6d. Issued 1969. Deleted 1970.　　　£5

M1810. Dunlop Footbridge Set.

Grey and white footbridge with 'Dunlop' decals and sign. Price 2/6d. Issued 1969. Deleted 1970.　　　£1

M1812. Starting Gate Set.

Fine automatic starting gate set consists of flags, signs and a fine building which is operated by a third person who acts as a track marshal. A Union Jack signals the the start of race and cuts in the power for a clean getaway. Price 12/6d. Issued 1969. Deleted 1970.　　　£2

M1816. Check-Point Bravo Set.

A Rally check-point and filling station in white, and green on fawn base with Shell petrol pumps and decals. The check-point building is yellow, blue and fawn. A clock-work motor raises and lowers the barrier across both tracks at variable pre-set intervals. This hazard can be used in various ways to make rally racing a real test of skill. Price 64/-. Issued 1969. Deleted 1970.　　　£7

M1817. Frontier Post Set.

Contains customs barrier and frontier post in yellow, green, black
and fawn, although colours may vary. A clock-work motor raises
and lowers the barrier across both tracks at variable pre-set
intervals. Price 55/-. Issued 1969. Deleted 1970.   £6

M1818. Mad Motorist Hazard Set.

One of the most hair-raising rally hazard ideas from Minic. A
hidden clock-work motor operates the Porsche non-electrical car
which backs out of a garage right across both tracks and disappears
back into the garage. In red and green with grey base and grey
doors, complete with two mechanical figures and two cars, one
green and one yellow, one mechanical and one electrical. Price 63/-.
Issued 1969. Deleted 1970.   £20

M1819. Pit Stop Set.

Contains a white, green, red and black building with six figures, a
left and right track junction and a straight single track. Price 17/6d.
Issued 1969. Deleted 1970.   £3

M1820. Double Pit Stop Set.

Contains two pit buildings to go on either side of a track, twelve
figures of mechanics, track marshal, onlookers etc., two left and
right track junctions and two straight single tracks. complete in
presentation box. Price 35/-. Issued 1969. Deleted 1970.   £7

TRI-ANG-MINIC PRE-WAR TRANSPORT VEHICLES

This is one of the most interesting sections in the whole Triang Minic range, especially as facts and figures about the pre-war releases were sometimes difficult to research. However, I have placed the following models in numerical order and it includes cars, vans, buses, tractors and bulldozers, not forgetting the odd caravan. There were only seven models which were first released in 1935 and although other models were made in the same year, they were held over and released after this time. The models all had strong reliable clockwork motors which were not only powerful but long running and were of a very strong construction. With care, these models would last for ever. The materials used in the production of these toys was tin plate, pressed steel and real rubber tyres. The liveries I give are the ones most commonly known, although I have included some of the rarer colour versions. There still could be other liveries in existence and if so I would like to hear from anyone who would like to correspond. These Minic clockwork toys were all to scale and were obtainable from only good toy shops and stores and they were made by Lines Brothers Limited at the Triang Works at Merton, London SW19 in England.

MODEL	MB	MU	GC
1M. Minic Ford £100 Saloon.			
Red with plated wheels and mudguards. Price 1/-. Issued 1936. Deleted 1940. 90mm.	£20	£15	£12
2M. Minic Ford Light Van.			
Blue or dark red. With 'Minic Transport Express (Triang Service)'. The first versions had white tyres and the latter changed to black. The former issue is worth double. Price 1/-. Issued 1936. Deleted 1940. 85mm.	£18	£15	£10
3M. Minic Ford Royal Mail Van.			
Red livery with 'Royal Mail' decals. Price 1/-. Issued 1936. Deleted 1940. 85mm.	£18	£15	£10
4M. Minic Sports Saloon.			
Ivory, new blue or primrose with plated mudguards, radiator, bumper and wheels and white or grey rubber tyres. Price 1/-. Issued 1935. Deleted 1940. 120mm.	£20	£15	£12
5M. Minic Limousine.			
Dark green with red plated mudguards. Also in new blue with ivory mudguards. Price 1/-. Issued 1935. Deleted 1940. 120mm.	£20	£15	£12
Also in red, rose-pink, grey or dark blue, otherwise as above.	£15	£12	£10
6M. Minic Cabriolet.			
Ivory and green with chrome mudguards, bumpers, grille and screen. Also in red with ivory mudguards, radiator and bumpers, chrome wheels and rubber tyres. Price 1/-. Issued 1935. Deleted 1940. 120mm.	£20	£15	£12
7M. Minic Town Coupe.			
Ivory or blue with chrome mudguards, wheels, bumpers and screen. Price 1/-. Issued 1936. Deleted 1940. 120mm.	£20	£15	£12
8M. Minic Open Sports Touring Car.			
Green with ivory hood or red with ivory hood and mud-guards. Also in ivory with red mudguards and hood. White tyres, chrome radiator and wheels. Price 1/-. Issued 1935. Deleted 1940. 120mm.	£20	£15	£12
9M. Minic Streamline Saloon Closed.			
Red, ivory or new blue, with chrome-plated bumpers, radiator, wheels with white rubber tyres. Price 1/-. Issued 1935. Deleted 1940. 126mm.	£18	£15	£12

10M. Minic Delivery Lorry.

Blue and stone or brown and grey, although other combinations of colours were produced. Black radiator, white tyres. Price 1/6d. Issued 1935. Deleted 1940. 140mm.

£25 £20 £1'

11M. Minic Tractor.

Green with red wheels or red with green wheels. With large white rubber tracks. Chrome steering wheels and radiator. Price 1/-. Issued 1935. Deleted 1940. 75mm.

£15 £12 £1(

12M. Minic Learner's Car.

Red or green with chrome bumpers, grille, wheels and screen. With white rubber tyres. Price 1/6d. Issued 1937. Deleted 1940. 118mm.

£20 £18 £1'

13M. Minic Racing Car.

Blue with white driver and the No.5 decal on rear. Grey or black rubber tyres with steel take off wheels. This was also made in red but is quite rare and worth treble. Price 1/6d. Issued 1936. Deleted 1940. 140mm.

£15 £12 £9

14M. Minic Streamline Open Saloon.

Ivory with red hood or red with ivory hood. With chromed grille, headlamps and screen and white rubber tyres. Price 1/-. Issued 1935. Deleted 1940. 126mm.

£18 £15 £12

Also in dark green or blue with a black hood, otherwise as above.

£14 £12 £9

15M. Minic Petrol Tank Lorry.

Green cab in various shades with tank, rear chassis and wings in red. In rare colour of stone worth treble. Tankers have either Shell or BP decals in gold lettering. Price 1/6d. Issued 1936. Deleted 1940. 145mm.

£25 £20 £15

16M. Minic Caravan. Non-Electric.

Two-tone green or maroon and stone which is worth double. The wire towing assembly is a two-pronged fixture made to hook over the rear bumper of a tin plate car. White tyres and chrome wheels. Price 1/-. Issued 1936. Deleted 1940. 116mm.

£12 £10 £7

17M. Minic Vauxhall Tourer.

This was one of the few Minic cars made with headlamps which were not part of the grille pressing. In fact they were very well made with bevelled edges and were fastened to the bar with a rivet. The early models are quite rare. Blue or green with chrome or black wings. In two-tone red and stone worth double. Most models have black or cream hood and white rubber tyres. Price 1/3d. Issued 1937. Deleted 1940. 120mm.

£20 £15 £12

18M. Minic Vauxhall Town Coupe.

Stone and green with black or cream hood. With chrome wheels, radiator, screen and headlights. Price 1/3d. Issued 1937. Deleted 1940. 120mm.

£20 £15 £12

19M. Minic Vauxhall Cabriolet.

This model had rounded wings, a petrol can, a luggage grid, a standard Vauxhall front and a Cabriolet back. After a while the petrol can and the luggage grid were deleted. In green livery, mainly because of the outbreak of war. Price 1/6d. Issued 1938. Deleted 1940. 120mm.

£20 £15 £12

MODEL	MB	MU	GC
0M. Minic Light Tank.			
Camouflage with dark green or grey tracks. Price 1/-. Issued 1935. Deleted 1940. 82mm.	£15	£12	£10
1M. Minic Triang Transport Van.			
Dark blue, dark green or red with chrome-plated wings, hubs and radiator and white rubber tyres. 'Minic Transport Road Rail Air and Sea Service' decals. Price 1/6d. Issued 1935. Deleted 1940. 140mm.	£18	£15	£12
2M. Minic Carter Paterson Van.			
Red cab and green box with 'Carter Paterson' decals, white tyres and chrome-plated wings, grille and radiator. Price 1/9d. Issued 1938. Deleted 1940. 140mm.	£20	£18	£15
3M. Minic Tipper Lorry.			
Red and green with white tyres and plated wings, hubs, grille and radiator. Price 1/3d. Issued 1936. Deleted 1940. 140mm.	£15	£12	£10
4M. Minic Luton Transport Van.			
In a two-tone combination of dark blue, grey, green and red with 'Minic Transport' decals. White tyres, chromed grille, radiator, wheels. Price 1/6d. Issued 1936. Deleted 1940. 140mm.	£18	£15	£10
With 'Atco Motor Mowers for the World's Finest Lawns'.	£20	£18	£15
4/AC. Military Minic Luton Van.			
In camouflage livery. Otherwise as above.	£15	£12	£10
4MCNZ. Minic New Zealand Luton Van.			
Dark blue and red with 'Winston Ltd Furniture Van' or 'Winston Tiles Makes Homes Beautiful'.	£25	£20	£15
5M. Minic Delivery Lorry with Cases.			
Red, green, blue or white with chrome hubs, grille, wings, white rubber tyres and six solid wooden block cases. Price 1/6d. Issued 1936. Deleted 1940. 140mm.	£20	£18	£15
6M. Minic Tractor and Trailer with Cases.			
Green with red wooden hubs and grey or white rubber tracks. Price 1/6d. Issued 1935. Deleted 1940. 185mm approx.	£25	£20	£15
7M. Minic Police Patrol Car.			
Blue with chromed wheels, grille, lights and radiator and white rubber tyres. With 'Police Patrol' decals on sides made in a limited supply in 1936. Price 1/6d. Deleted 1940. 134mm.	£35	£30	£25
8M. Minic Sports Tourer with Boat on Trailer.			
Red with matching red boat on black or grey trailer. White rubber tyres and chromed hubs, radiator etc. Price 3/6d. Issued 1936. Deleted 1940. 238mm.	£35	£30	£25
9M. Traffic Control Car.			
Dark blue with black wings, chromed wings and radiator and black tyres. Two policemen cast in. Price 1/9d. Issued 1938. Deleted 1940. 126mm.	£15	£12	£9
10M. Minic Mechanical Horse and Pantechnicon.			
Blue or fawn tractor with red box on rear, white tyres, no adverts. Price 2/-. Issued 1935. Deleted 1940. 195mm.	£20	£18	£15

31M. Minic Mechanical Horse & Fuel Oil Tanker.

Green cab, red tanker body and chrome wings, radiator and wheels. 'Shell & B.P. Fuel Oil' decals. Price 2/6d. Issued 1936. Deleted 1940. 178mm.

£25   £20   £1'

32M. Minic Dust Cart.

Green cab and red cart body with chromed sliding doors, wings and radiator and white tyres. 'Triang-Minic' decals. Price 2/3d. Issued 1936. Deleted 1940. 140mm.

£18   £15   £1(

33M. Minic Steam Roller.

Dark green or dark red body with brass funnel and front axle swivel unit. Bright tin plate cylinder box and push-rod. Price 1/-. Issued 1935. Deleted 1940. 140mm.

£16   £14   £1(

34M. Minic Tourer with Passengers.

Green, dark blue or yellow body with chromed wings, radiator, wheels etc. A chauffeur with a green uniform and silver buttons in driving seat. Next to him is a little girl in a yellow dress. Man in rear of car has a dark blue suit and a trilby. Woman in rear wears a blue dress. Price 3/6d. Issued 1938. Deleted 1940. 126mm.

£25   £20   £15

35M. Minic Rolls Type Tourer (Non-Electric).

White or black and red worth double. Opening boot with all chromed wings, radiator and headlights. White tyres. Price 2/6d. Issued 1937. Deleted 1940. 134mm.

£18   £15   £10

36M. Minic Daimler Type Tourer, Non-Electric.

White with all chromed parts. White tyres. Price 2/6d. Issued 1937. Deleted 1940. 134mm.

£15   £12   £10

37M. Minic Bentley Type Tourer.

White with opening boot and all chrome radiator, grille etc. Also in red and black with white tyres. Price 2/6d. Issued 1937. Deleted 1940. 134mm.

£18   £15   £12

38M. Minic Caravan Set.

Two-tone green and red and cream with all chrome parts. Price 3/11d. Issued 1936. Deleted 1940. 230mm.

£25   £20   £15

39M. Minic Taxi.

Two-tone green with varying shades, black hood and chassis with luggage compartment in stone colour. With spare wheel and chromed wheels. Number plate reads 'LBL 174. Price 2/6d. Issued 1938. Deleted 1940. 106mm.

£25   £20   £15

40M. Minic Mechanical Horse & Trailer with Cases.

Red, blue or green with chrome wings, radiator, screen and wooden cases. Lights. Price 2/6d. Issued 1938. Deleted 1940. 188mm.

£25   £20   £15

MODEL	MB	MU	GC

1M. Minic Caravan with Electric Light & Battery.
Two-tone green or maroon and stone with brass battery connector.
Price 2/-. Issued 1936. Deleted 1940. 108mm.

	£18	£15	£10

2M. Minic Rolls Type Sedanca, Non-Electric.
Off-white or cream which is worth double. With chromed wheels,
radiator, wheels etc., and white tyres. Price 2/6d. Issued 1938.
Deleted 1940. 128mm.

	£16	£14	£12

3M. Minic Daimler Type Sedanca. Non-Electric.
White or cream with all chromed parts and white tyres. Price 2/6d.
Issued 1938. Deleted 1940. 128mm.

	£18	£15	£12

4M. Minic Traction Engine.
Green or red which is worth double. With brass smoke box door
and black wheels. Price 2/6d. Issued 1938. Deleted 1940. 120mm.

	£30	£25	£20

5M. Minic Bentley Type Sunshine Saloon. Non-Electric.
Cream, stone or white with chrome radiator, wheels, wings etc.
Price 2/6d. Issued 1938. Deleted 1940. 134mm.

	£18	£15	£12

6M. Minic Daimler Type Sunshine Saloon. Non-Electric.
White or black which is worth double. With chrome radiator,
wheels, wings etc. Price 2/6d. Issued 1938. Deleted 1940. 134mm.

	£18	£15	£12

7M. Minic Rolls Type Sunshine Saloon. Non-Electric.
White or rose pink which is worth double. With chrome radiator,
wheels, wings etc. Price 2/6d. Issued 1938. Deleted 1940. 134mm.

	£18	£15	£12

8M. Minic Breakdown Lorry with Mechanical Crane.
Green and brown. Price 1/6d. Issued 1936. Deleted 1940. 140mm.

	£20	£18	£15

9ME. Minic Searchlight Lorry.
Cab and searchlight are dark green and rear flat bed is red. Battery
box is green or red and the model has chromed wings, radiator and
reflector. Price 2/3d. Issued 1936. Deleted 1940. 140mm.

	£25	£20	£15

9MECF. Minic Searchlight Lorry.
A rare camouflaged livery made in a limited supply in 1936. Price
2/3d. Deleted 1940. 140mm.

	£50	£45	£35

MODEL	MB	MU	G
50ME. Rolls-Type Sedanca with Electric			
White or stone pink with chromed parts and battery. Complete with small tear-drop bulbs in chromed headlights. Price 2/3d. Issued 1937. Deleted 1940. 128mm.	£18	£15	£1:
51ME. Minic Daimler Type Sedanca with Electric.			
White, cream or stone pink with chromed parts including headlights with small tear-drop bulb. Complete with battery. Price 2/3d. Issued 1937. Deleted 1940. 128mm.	£18	£15	£1:
52M. Minic Single Decker Bus.			
Red with various adverts and destination decals. Price 2/6d. Issued 1936. Deleted 1940. 186mm.	£30	£25	£2(
53M. Minic Single Decker Bus.			
Green with various adverts and destination decals. Price 2/6d. Issued 1936. Deleted 1940. 186mm.	£20	£18	£15
54M. Minic Traction Engine and Trailer with Cases.			
Green and red with black wheels. Price 3/6d. Issued 1937. Deleted 1940. 222mm.	£30	£25	£2C
55ME. Minic Bentley Type Tourer With Electric.			
White. Complete with battery and all chromed parts. Price 3/-. Issued 1937. Deleted 1940. 134mm.	£20	£18	£15
56ME. Minic Rolls Type Sunshine Saloon with Electric.			
White or fawn. Complete with battery and all chromed parts. Price 3/-. Issued 1937. Deleted 1940. 134mm.	£20	£18	£15
57ME. Minic Bentley Type Sunshine Saloon with Electric.			
White or black and white, complete with battery and all chromed parts. Price 3/-. Issued 1937. Deleted 1940. 134mm.	£20	£18	£15
58ME. Minic Daimler Type Sunshine Saloon with Electric.			
White, stone livery or rose pink which is worth double. Complete with battery and all chromed parts. Price 3/-. Issued 1937. Deleted 1940. 134mm.	£20	£18	£15
59ME. Minic Caravan Set.			
Touring caravan with limousine and passengers. The details are exactly as 34M and 41M. Price 3/11d. Issued 1938. Deleted 1940. 245mm.	£25	£20	£15
60M. Minic Double-Decker Bus.			
Red with several adverts, including 'Drink Delicious Ovaltine for Health' or 'Bovril' or 'Bisto'. Price 2/11d. Issued 1935. Deleted 1940. 184mm.	£30	£25	£20
61M. Minic Double Decker Bus.			
Green with several adverts, including 'Drink Delicious Ovaltine for Health' or 'Bovril' or 'Bisto'. Price 2/11d. Issued 1935. Deleted 1940. 184mm.	£35	£30	£25
62M. Minic Fire Engine.			
Red with red ladders. Price 3/-. Issued 1937. Deleted 1940. 158mm.	£20	£18	£15
62ME. Minic Fire Engine with Electric Headlamps and Battery.			
Red with red ladders. Price 5/-. Issued 1937. Deleted 1940. 158mm.	£25	£20	£15

MODEL	MB	MU	GC

3M. Minic Set No.1.

rare pre-war set containing a limousine, a Cabriolet, a Ford £100 saloon, a sports saloon and an open tourer in special presentation box. Price 7/6d. Issued 1936. Deleted 1940.

	£50	£45	£40

Minic 64M. Minic Set No.2.

ontains a petrol tank lorry, a Carter Patterson lorry, a Triang ransport van, a tipper lorry, a racing car, a light tank, a Ford 100 saloon, a petrol tank wagon, a limousine and a Ford light van. rice 15/6d. Issued 1936. Deleted 1940.

	£125	£100	£75

5M. Minic Construction Set.

nother rare set. A child could spend many hours making up six fferent vehicles which included a tractor, a delivery van, a reamlined tourer, a Cabriolet limousine and a transport van. Price 5/-. Issued 1936. Deleted 1940.

	£50	£45	£35

5M. Minic Six Wheel Army Lorry.

ark green or camouflage. Price 2/6d. Issued 1939. Deleted 1940. 40mm. Price for dark green model.

	£20	£18	£15

rice for camouflage model

	£25	£20	£15

7M. Minic Farm Lorry.

lue cab with rear colours that vary from red, green, black and one. Complete with chromed or black wings. Price 2/-. Issued 39. Deleted 1940. 140mm.

	£18	£15	£12

8M. Minic Timber Lorry.

lue and green with four wooden planks. Price 2/-. Issued 1939. eleted 1940. 140mm.

	£18	£15	£12

9M. Minic Canvas Tilt Lorry.

ark green with fawn or brown fabric cover. Price 2/-. Issued 1939. eleted 1940. 140mm.

	£18	£15	£12

9MCF. Minic Canvas Tilt Lorry in Camouflage.

amouflage. Price 2/3d. Issued 1939. Deleted 1940. 140mm.

	£25	£20	£15

9M. Minic Coal Lorry.

nother very neat and rare model with a green or blue cab and a d or black rear wagon. With 'Minic Coal Deliveries' decal. Price 3d. Issued 1939. Deleted 1940. 152mm.

	£50	£45	£35

.M. Minic Mechanical Horse and Milk Tanker.

rey, white or cream with black tyres. Complete with 'Minic airies' decals. Price 2/6d. Issued 1939. Deleted 1940. 178mm.

	£25	£20	£15

.M. Minic Mechanical Horse and Lorry with Barrels.

ed cab, green body and brown barrels with chrome or black ings. Price 2/6d. Issued 1939. Deleted 1940. 178mm.

	£25	£20	£15

3M. Minic Cable Lorry.

ed or green cab and green trailer with cable decals. Price 2/6d. sued 1939. Deleted 1940. 210mm.

	£25	£20	£15

.M. Minic Log Lorry.

ed or green cab with blue trailer. Price 2/6d. Issued 1939. Deleted 40. 202mm.

	£25	£20	£15

MODEL	MB	MU	G
75M. Minic Ambulance.			
White with a red cross on a white background or a white cross on a red background which is a mistake and is worth treble. This model has cast wheels and two long green benches in the back. Price 1/11d. Issued 1938. Deleted 1940. 144mm.	£20	£15	£1
76M. Minic Balloon Barrage Wagon/Trailer.			
This is definitely the rarest Minic model ever made. The balloon itself was actually a second Minic Field Kit made of stiffened fabric painted silver and was supplied in three main pieces which then had to be sewn together. The ring to which the twelve cables are fixed is of white plastic. The balloon has an instruction leaflet in its box. The balloon is 330mm long and the lorry and trailer (240mm long) are in camouflage and at a very reasonable price of 3/6d. It was made in 1939 and disappeared almost immediatley because of the war. Complete with a clockwork winch mechanism on the lorry and a gas cylinder on the trailer.	£250		
77M. Minic Double Decker Trolley Bus.			
Red or green the latter being worth treble and very rare. Complete with 'Bovril' advert. Price 3/6d. Issued 1939. Deleted 1940. 178mm.	£35	£30	£2
78M. Minic Jeep.			
Green or camouflage. Price 1/9d. Issued 1939. Deleted 1940. 84mm.	£10	£7	£5
79M. Minic G.W.R. Van.			
Brown and cream with 'Great Western Railway' and 'Express Cartage Service'. Decals. Made in 1939 and deleted in 1940. Re-issued after the War. Beware of copies. Determine the authentic model with the advice of an expert. Price 2/-. 148mm.	£30	£25	£2
80M. Minic L.M.S. Van.			
Maroon and black with 'Express Parcels' decals. Price 2/-. Issued 1939. Deleted 1940. 148mm.	£25	£20	£1
81M. Minic L.N.E.R. Van.			
Dark blue and gold with 'Express Parcel Service' decals. Price 2/-. Issued 1939. Deleted 1940. 148mm.	£25	£20	£1
82M. Minic S.R. Van.			
Dark green with 'Triang Southern Railway' decals. Price 2/-. Issued 1939. Deleted 1940. 148mm.	£30	£25	£2
83M. Minic Farm Tractor			
Green or red. Price 1/11d. Issued 1938. Deleted 1940. 90mm.	£12	£9	£7
84M. Minic Jeep			
Dark Green or camouflage. Price 1/11. Issued 1938. Deleted 1940. 160mm.	£12	£9	£7

PRE-WAR SPECIALS

Apart from the clockwork Minic models and the plastic friction-driven models, the firm of Triang introduced some large trucks, cranes and tanks which were pull-or-push-along playthings. These pre-war specials were amongst the best ever made.

If the children could not sit inside the actual toy they could sit on top of many of them and have hours of endless fun. While the trucks carried soil, bricks and sand, the large double decker bus would carry an excited child quite safely on top of it.

Many of these models came in two sizes, medium and very large. The main colours were red and blue with the occasional green. While some models were 612mm in length and approximately 306mm in height, the smaller ones were approximately 306mm in length and 204mm in height.

MODEL	GC	FC
Boy on a Swing.		
Tin plate with strong compact clockwork motor. Celluloid figure on wire clips which cleverly does acrobatic movements with astonishing accuracy. Price 7/6d. Issued 1933. Deleted 1940. 190mm approx. height.	£100	£75

MODEL	GC	FC
Jack in the Boat.		
Green boat with brown oars and tin plate figure. Brown interior. Clockwork. Clever movement. Price 7/6d. Issued 1933. Deleted 1939. 208mm approx.	£100	£75
The Dancing Clown.		
Tin plate. White with red and blue dots and black buttons. This clown figure stands on a box attached to a wire and when lever is released clown dances around a green pole. Price 3/11d. Issued 1933. Deleted 1940. 158mm approx. height when attached.	£80	£60
Triang Nippy Tractor No.1.		
Dark green, brown or red livery with thick rubber tracks. Complete with driver and powerful clockwork motor. Price 1/-. Issued 1933. Deleted 1940. 134mm.	£30	£20
Triang Nippy Tractor No.2.		
Steel construction in red-painted cellulose with a reliable motor, control lever and driver. With thick rubber bands on steel wheels. Price 3/11d. Issued 1933. Deleted 1940. 216mm.	£50	£35
Whippet Climbing Tank No.1.		
Dark green or camouflage with strong motor and control lever with swivelling gun turret, steel wheels and rubber bands. Price 1/-. Issued 1933. Deleted 1940. 134mm.	£35	£25

Tiger Climbing and Fighting Tank No.2.

Dark green or camouflage. Made of solid steel with wide rubber bands, swivelling gun turret with firing device and control lever. Price 5/11d. Issued 1933. Deleted 1940. 255mm.   £75   £6

Magic Sports Car.

Red, green or stone livery. This car does several tricks at the pressing of a lever. Made of steel with a powerful motor. Price 3/11d. Issued 1933. Deleted 1940. 190mm.   £75   £6

Large Sports Saloon Car.

Blue, red or black. This is made of all steel with a powerful clockwork motor that gives the sound of a racing sports car which is quite authentic. Steel wheels and thick rubber tyres enables it to travel on any surface. Complete with driver in red or white uniform. Price 3/6d. Issued 1933. Deleted 1940. 200mm.   £100   £7

Steel Tricycle.

Red with black pedals mounted on large front wheel, two smaller wheels at the rear with thick grey rubber tyres. Price 21/-. Issued 1936. Deleted 1939. 84cms.   £30   £2

Child's Mangle.

This was advertised with the slogan, 'Just Like Mummy's'. With wooden rollers and triple cogged wheels on green or red painted trestle base which stood on smaller wheels. This model could actually wring out many items in a similar way that the larger models could. Price 12/6d. Issued 1935. Deleted 1939. 51cms. approx. width.   £35   £3

Wooden Horse and Cart.

Finely painted in red, white and blue, complete with harness and tinkle bells. Could be pulled or pushed. Price 4/11d. Issued 1936. Deleted 1939. 29cm High.   £15   £1

The Puff Puff Train.

Blue with black funnel, blue wheels and rubber tyres. When the train is pulled along, the sound 'Puff-Puff' can be heard plainly and loudly. Very realistic toy. Price 10/6d. Issued 1936. Deleted 1939. 40cm approx. length; 18cm approx. height.   £20   £1

Battery Driven Bus.

Red with black and gold lines. Driver's seat in red with black grille and silver bumpers. 'London, Glasgow, Edinburgh' decals. Room for six passengers. Seats turn up at rear to make a fine luggage rack. Thick rubber tyres and headlights and tail light. Spare wheel fitted in compartment above driver's head or fixed on at rear. Price £45. Issued 1930. Deleted 1940. 3ft long; 22in wide.   £650   £5

Battery Driven A/C Special Car.

This child's car caught the eye of every man who had money and loved his children. A winner all the way when pedals were replaced by a battery-powered car. In gold with black and silver lines. Black mudguards and brass step at foot of opening door. Silver grille, metal bumpers and black number plate. Silver Eagle or wings motif and statuette on nose of bonnet. Gold steering wheel, spoked wheels and solid rubber tyres. Rear seat and luggage rack, horn, reflectors and speed of five miles per hour for child's safety, although models were known to be faster. They were sold only on

MODEL	GC	FC

the understanding that children would drive them under parental or guardian supervision. Price £47. Issued 1929. Deleted 1940. 2ft 9in long; 22in wide.

£500 £400

No.7. Grand Prix Racing Car.

Dark blue with 'No.7' decal on rear. Metal grille and wheels with rubber tyres. Driver with red helmet. Price 3/6d. Issued 1938. Deleted 1940. 127mm.

£75 £50

Minic Flying Boat.

Silver with red lines and passengers faces painted on windows. Painted pilot and crew. Price 5/11d. Issued 1938. Deleted 1940. 104mm approx.

£65 £55

Minic Yacht with Sail.

Tin plate with strong clockwork motor. Boat also has white linen sail. Boat is red and white. Price 7/6d. Issued 1938. Deleted 1940. 154mm.

£75 £65

Wolseley Police Car with Siren.

Black with metal wheels and grille and rubber tyres. Police sign on roof, mudguards and two police tin-plate figures, driver and passenger. Clockwork motor. Price 8/11d. Issued 1938. Deleted 1940. 199mm approx.

£100 £80

Petrol Tanker 'Shell'.

Clockwork model in red, green and yellow lettering. Metal wheels with black rubber tyres. Metal grille and bumpers with black mudguards. Price 8/11d. Issued 1938. Deleted 1940. 141mm approx.

£50 £40

Wooden Rocking Horse.

Dapple-grey body with black or brown applied harness. Fixed on adaptable stand so horse could be fixed to rockers or to wheels. A clever first attempt by Lines Brothers for a dual-purpose plaything. Price 99/-. Issued 1937. Deleted 1939. 163cm.

£250 £200

Large Triang Play Wagon.

This was one of the first delivery vans made of all steel with strong rubber wheels. A child could easily ride on top of these adaptable wagons. Several adverts were connected with these toys, two of which were 'Fry's Milk Chocolate' and 'Fry's Pure Concentrated Cocoa' in the authentic colours of the firm of that period. The Fry's advert showed a picture of a girl holding a cocoa tin. The other advert connected with this model was 'Pears Soap', with a picture of a large bar of soap and a baby crawling. Price 39/-. Issued 1936. Deleted 1939. 46cms approx. length; 25cms high.

£75 £65

The Safety Scooter.

Perfectly safe but also provides plenty of excitement and healthy exercise. One foot is placed on the platform while the other is used to strike the ground. When sufficient speed is obtained, both feet can be placed and balanced on the platform. Some models had seats but these were seldom seen and were eventually deleted. In red and black or brown and black, with tin plate and rubber wheels. price complete with pillar and seat 10/6d. Without seat and pillar 7/11d. Issued 1929. Deleted 1936. Length 91 cm approx; width 20cm approx; height 78cms.

£350 £300

Triang Pedal Yacht.

White, blue and red with pneumatic tyres and black pedals. A novel car that can obtain high speeds when red sails are raised, an idea which developed in the American town of Henderson, where cars were often fitted with sails and raced along the sands. Meant for the older boy or girl, with parental supervision. Price 50/-. Issued 1930. Deleted 1939. Approx. length 127cms, approx. width 77cms, approx. height when sail is raised 100cms. £500 £3

Baby Austin. Saloon.

Blue and black with pneumatic tyres, whited-walled and ribbed. With Honky Horn, petrol can, spare wheel on side and working tail and headlights. Price £7. Issued 1931. Deleted 1939. Approx. length 91cms, width 66cms, and height approx. 100cms. This was one of the strongest model cars ever built. £750 £5

Pedal Car Number 1.

In red with all action peddle mechanism, horn, imitation lights and mudguards. Price 25/-. Issued 1931. Deleted 1939. Length 75cms, width 50cms, height 45cms. £100 £7

Pedal Car Number 2.

Red and black or blue and black with horn, mudguards, running boards and lights. Solid wheels. Price 35/-. Issued 1932. Deleted 1939. Approx length 85cms, width 50cms, height approx 30cms. £150 £1

Pedal Car Number 3.

Blue and black with all battery-operated lights, petrol can on running boards, mudguards, horn and steel driving mirror. Thick rubber tyres on spoked wheels. Price 45/-. Issued 1933. Deleted 1939. Approx. length 85cms, width 50cms approx., height 56cms. £250 £1

Mickey and Donald's Rail Cart.

Green and red with metal wheels and clockwork motor. Price 2/6d. Issued 1936. Deleted 1939. 18cms. £25 £2

Clockwork Rowing Boat.

Red and cream, complete with oars, spare key and rubber plug for keyhole. Travels at a very fast speed along the water, reliable springs. Price 3/6d. Issued 1936. Deleted 1939. 18cms approx. £25 £2

Triang Arcitex.

Scale model construction set in special presentation box. Original Price 3/6d. Issued 1935. Deleted 1939. £20 £1

Bread Cart.

A flywheel-driven tin-plate clockwork walking girl pulling a bread cart. The girl walks on a revolving wheel beneath her legs. In brown and white with 'Bero Flour' advert. Price 7/6d. Issued 1936. Deleted 1939. 18cm approx. £35 £3

Horse and Cart.

Orange, brown and dark yellow. This is a clockwork tin-plate horse and cart with drive wheel. Price 3/6d. Issued 1935. Deleted 1939. 224cm approx. £50 £4

Wooden Rocking Horse.

Dark brown mounted on an oak swinging stand. Price £2.10d. Issued 1935. Deleted 1939. 126cm approx. £40 £3

MODEL	GC	FC

Dolls' House.

A mock Tudor timbered facade with shuttered and curtained
windows. Opens up into four sections to reveal five rooms, staircase
and double garage. This fine toy includes a bath, stove, welsh
dresser, sink, sundial and three fireplaces. Cream blue and white
mounted on an oak chest with two large drawers. Price £4.10s.
Issued 1934. Deleted 1939. 120cms wide x 99cms high approx.   £150   £100

No.2. Dolls' House.

In simulated brick and pebble dash design, often added to after
original manufacture by handymen. Five shuttered windows in
white and red with matching hinged front door. Front of the house
slides out to reveal four rooms, staircase, bathroom and kitchen.
The house contains wooden furniture, hanging pictures and a
piano. Fully fitted bathroom, cooking range and accessories. Three
small dolls, mother, father and little girl, the largest being 7.5cms.
Clocks, carpet-sweeper, iron on stand, pots, pans, and a baby's
pram containing small baby. A unique accessory is the tiny baby's
rattle tied with a pink ribbon to the pram. This must be one of the
smallest toy items in the world, and very rare. Original price
£12-15-6d. Issued 1934. Deleted 1939. 66cms wide and 75cms high.   £250   £200

Set of Dolls' House Furniture.

In a special Triang presentation box, including a dresser with shelf
supports and handles, upright piano, wardrobe, dressing table, stool
and trinket box. Wooden rocking chair, upholstered chairs, dining
table, chests, wash-hand stand, poss-tub, mangle and clothes basket.
Work-table of oak and metal, with pictures and small ornaments.
Price 50/-. Issued 1934. Deleted 1939. Box size approx. 46cms x
61cms and 20 cms deep.   £50   £45

A Toy Fort.

Cream, brown and dark red with towers, drawbridge moat and
rocky hill. Complete with soldiers that fit into bottom of fort. Price
17/6d. Issued 1935. Deleted 1939. 20cms high, 38cms wide.   £30   £20

Train Set.

Containing an 0-4-0 Hurricane Locomotive; tender; three passenger
coaches and six pieces of curved track in an original 'Alan
Anderson' box. Green, black and gold or red, black and gold. Price
7/6d. Issued 1935. Deleted 1939.   £30   £25

Metal Rocking Horse.

The moulded details in red and white with brown saddle include
mane and tail with foot rests and two handles protruding from the
side of animal's head and mounted on two curved metal rockers.
Price 21/-. Issued 1935. Deleted 1939. 109 cms wide.   £40   £25

Push-A-Long Wooden Horse.

With turned wooden handle in red, cream and black, imitation
leather trimmings, white mane and horsehair tail. Mounted on
wooden base, raised on four metal wheels. Price 25/-. Issued 1935.
Deleted 1939. 66cms.   £50   £45

Wooden Dolls House No.3.

Blue, white and golden yellow. The facade is decorated with mock
timber and painted flowers. It opens into two sections, revealing
four rooms, staircase, garage and part kitchen and wash-house.
Assorted wooden furniture, a die-cast fireplace, grandfather clock,
bathroom suite, kitchen utensils and eight wooden and cloth dolls.
Price 95/-. Issued 1936. Deleted 1939. 68cms wide, 41cms high.   £75   £50

Two-Storey Dolls House.

Red, cream and white with a green fence and a brick design, large
upstairs window, one downstairs window and front door, all with
arched pediments. Opens to reveal two rooms, kitchen, garage and
bathroom. With sofa, chairs, piano, wardrobe, two fire places and
enamel tea set on table. Also garden, greenhouse, lawn-mower and
several tools. Price £7.10s. Issued 1936. Deleted 1939. 56cms high,
on large baseboard. £200 £150

Dolls Tea Set & Pastry Set.

More than 50 pieces made of tin with nursery rhyme designs. Blue,
white and pink with gilt and painted floral design on pastry set.
Rolling pins, pastry cutters and bowls in original nursery rhyme
presentation box. Price 57/6d. Issued 1936. Deleted 1939. Box
46cms x 61 cms x 15cms deep. £55 £50

A Doll's Bed.

Oak bed complete with head and tail boards and base of interwoven
metal strips. Complete with mattress, bolster pillow, two sheets,
blanket and bedcover. Price 65/-. Issued 1936. Deleted 1939. 67cms
long, 42 cms wide. £40 £35

A Doll's Cot.

Pale blue wooden parts (five), complete with assorted bedding.
Hand-painted animal pictures on sides, front and rear. Price 7/6d.
Issued 1936. Deleted 1939. 64cms x 38cms x 49cms. £20 £15

A Doll's High Chair.

Blue with real leather straps, and beads for counting. Also available
in dark or medium oak for 21/-, otherwise as above. Price 10/6d.
Issued 1936. Deleted 1939. 25cms x 64cms. £15 £10

Charlie McCarthy Ventriloquists Doll.

With celluloid head and articulated mouth, painted features,
monocle and jointed rag body. Hands and feet of moulded celluloid.
Doll is dressed in an immaculate dinner suit. With matching top
hat. Price 45/-. Issued 1936. Deleted 1939. 54cms approx. £75 £65

Small Wooden Charlie McCarthy Ventriloquists Doll.

Wood with painted face, dressed in evening suit with peaked cap.
Strings in back and head on stick. Price 17/6d. Issued 1936.
Deleted 1939. 45cms approx. £30 £25

Junior Doll's Pram.

A fine combination of steel and leather and one of the better
quality Triang products. Blue and black with stainless steel fittings.
With darker blue pram cover, large spoked wheels on reliable
springs, complete with black mudguards. Folding hood and rubber
tyres. One of the first prams with ball-bearing wheels. Price
£7.17.6d. Issued 1936. Deleted 1939. 87cms. £75 £65

Clockwork Sports Car.

This fine vehicle is in red with blue seat or in blue with red seat.
Mudguards match seat. With white rubber tyres and hand brake on
side. Complete with driver and windshield. In special presentation
box. Price 5/6d. Issued 1936. Deleted 1939. 31cms. £75 £65

Speedboat Racer.

Blue, white and silver, finished in cream, red and gold. With a
clockwork mechanism concealed within. Stopper cork to keep water

MODEL	GC	FC

ut, fixes in key slot. In box complete with a spare key. Price
4/11d. Issued 1936. Deleted 1939. 45cms.   £75   £65

Pull Along Horse and Cart.

White and grey with hide, mane, tail, simulated leather tack.
Mounted on wooden base. Cart-wheels and wheel under horse can
be detached for transporting. Complete with box. Price 37/6d.
Issued 1935. Deleted 1939. 61 x 33cms.   £40   £35

Neat Prancing Grey Rocking Horse.

Grey with head slightly to the right, flared nostrils, glass eyes
(inset), leather saddle, horsehair mane and long tail. Mounted on
strong rockers. Stirrups adjustable. Complete with whip and riding
ap. Price £10. Issued 1936. Deleted 1939. 114 x 196cms.   £350   £300

Gyro Cycle.

A tin plate cycle with a red frame and red and white wheels
carrying the figure of a boy in red, yellow and orange with a red
cap and black shoes. A unique toy worked by means of a string and
gyro balance mechanism. It was essential that once this toy was
activated it should not be picked up or stopped until it had run its
course. Made in 1939 and deleted almost at once. Copies were
made post-war but not anywhere near the same quality. Price 7/6d.
108mm approx.   £50   £40

Minic Fire Station.

Cream and red with opening doors complete with a red fire engine
which was fully wound up before being placed inside, and was
released automatically. Also a working bell system run on a battery.
Price 15/-. Issued 1939. Deleted 1940. 250 x 250 x 250mm. £75   £50

Minic Service Station.

Red and cream with one sliding shutter door and three petrol
pumps with 'Shell' decals. Complete with two cars and a
breakdown wagon. The colours of vehicles vary between blue,
green, grey and red. Made in 1939 and deleted in 1940. Again,
copies of this were made after the War in the 1950s. The quality
and design however were not the same. The wooden structure of
the garage pre-war was replaced by heavy laminated card post-war,
although the colours remained almost the same. Price 10/6d.
250mm approx.   £50   £40

Triang Wooden Doll's House.

The facade is decorated with mock timber and painted flowers and
opens in two sections to reveal four rooms, staircase and garage.
House includes assortment of wooden furniture, a diecast fireplace,
grandfather clock, bathroom suite, kitchen utensils, eight small
wooden and cloth dolls and a unique library with miniature books
etc. A pipe is on a mantle piece and the set of fireside items is
quite rare. Tidy, poker, brush set and fender all in copper and
brass. Price £7.10s. Issued 1936. Deleted 1940. 680 x 410mm.   £500   £4

Clockwork Chicken and Chick.

A fine clockwork toy in yellow and white or cream and blue. After
being wound up, the chicken runs on a wheel mechanism and the
chick follows. Price 3/6d. Issued 1939. Deleted 1940.   £20   £1

Clockwork Train.

Red body with black and white lineage. This is another first-class
tin-plate toy. Price 2/6d. Issued 1939. Deleted 1940. 210mm.
approx.   £15   £1(

Triang Motorcyclist.

Red and cream. All tin plate with a strong clockwork mechanism
and driving rear wheel. Figure is in black, red and blue. Price 2/6d.
Issued 1939. Deleted 1940. 215mm.   £30   £2

Twin Police Motorcyclists.

Tin plate, clockwork dual mechanism. Cream and red bikes with
police decals and blue riders. Price 3/6d. Issued 1939. Deleted
1940. 215mm.   £50   £4

POST-WAR PUSH, PEDAL AND GO TOYS

n the post-war years, apart from the clockwork Minic models, there were several large trucks,
cranes, tanks and other toys that a child could pull or push along with comparative ease. Some
of the most outstanding models in this new range were the railway engines in red, blue and the
rarer colour of green. Even when children could not sit inside the actual toy, they could sit on
top of the plaything and travel at fast speeds. Some of these trucks and toys had handles for a
child to grip. The trucks carried soil, bricks and sand for the boys, the girls were quite happy
with their pram or fairy cycle. Although the models varied in size, they were approximately
612mm long and 306mm high. The smaller models were approximately 306mm in length and
204mm in height. Most models were made of steel and had steel wheels and rubber tyres. The
est were made of wood and often had steel wheels.

MODEL	GC	FC
Triang Express Delivery Van (Large).		
Red body with black mudguards and running board, red wheels and black tyres. Headlights, number plates and bumpers. 'Triang Express Parcels Delivery' decals in yellow or orange lettering. Price 49/11d. Issued 1959. Deleted 1970. 612 x 306mm.	£25	£15
Medium delivery van. Price 10/6d. 306 x 204mm. Otherwise as above.	£12	£9
Triang Puff-Puff Engine (Large).		
Red. Has matching wheels with black rubber section that makes a realistic 'puff-puff' sound as the model is pulled along. Children can sit on it and push themselves along with their feet. With yellow or black chimney. Price 59/11d. Issued 1960. Deleted 1970. 612 x 306mm	£25	£15
Medium Puff-Puff Engine. Price 19/6d. 306 x 204mm. Otherwise as above.	£12	£9
Triang Puff-Puff Model (Large).		
Blue with matching wheels and black funnel or stack. Price 59/11d. Issued 1960. Deleted 1970. 612 x 306mm.	£15	£12
Medium Puff-Puff Engine. Price 19/6d. 306 x 204mm.	£10	£8
Triang Puff-Puff Model (Large).		
Green with black stack and green wheels. Rare colour. Price 59/11d. Issued 1961. Deleted 1970. 612 x 306mm.	£50	£40
Triang Puff-Model (medium). Price 19/6d. 306 x 204mm. Otherwise as above.	£35	£30
Triang Land Rover and Trailer (Large).		
Fawn with cream line trim. Has matching trailer and driver. Price 56/11d. Issued 1962. Deleted 1970. 816mm.	£15	£12
Silver Link Airliner.		
Tin plate, silver, painted windows, door and rubber wheels. Clockwork, runs a fair distance on floor. Painted figure pilot and 22 passengers. Red or brown props, twin engine. Price 10/6d. Issued 1953. Deleted 1970. 375mm approx.	£100	£75
Pickford's Removal Van.		
Dark green with wording in white. Not clockwork. Rubber wheels, opening rear doors. Lights with battery. Driver and mate in cab with black interior. Price 15/11d. Issued 1952. Deleted 1970. 350 x 175mm approx.	£250	£200
Triang Jeep.		
Cream with black steering wheel and seats. Price 95/- Issued 1952. Deleted 1970. 355mm.	£50	£45

MODEL	GC	F
Triang Pedal Car.		
Red body, black steering wheel and seat and black or red wheels. With black wings. Price 45/-. Issued 1952. Deleted 1970. 915 x 355mm.	£75	£6
With black body and red wings. Rare.	£250	£2
Triang Pedal Car.		
Red with black mudguards and steering wheel, red wheels and thick rubber tyres. Price 32/6d. Issued 1950. Deleted 1970. 900 x 350mm approx.	£75	£5
Triang Doll's Pram.		
Black, blue or red with spoked wheels and real rubber tyres, complete with folding hood and covers. The shades and colours may vary. Price 55/-. Issued 1950. Deleted 1970. 850 x 350mm approx.	£50	£4
Triang Fairy Cycle.		
Red, black or blue frames with spoked metal wheels and rubber tyres. Unique and sturdy. Price 75/-. Issued 1951. Deleted 1970. 900 x 400mm approx.	£40	£3
Triang Tricycle.		
Red, blue or black. With strong spoked wheels and rubber tyres. Price 75/-. Issued 1952. Deleted 1970. 800 x 300mm approx.	£40	£3
Triang Tipper Truck.		
Green cab and red tipper body, although colours were often reversed. Made of strong steel. Price 21/-. Issued 1950. Deleted 1965. 350mm approx.	£25	£2
Minic Charabanc (Wood).		
Cream and red or cream and dark brown with opening rear doors. Steel wheels and rubber tyres. Price 25/-. Issued 1950. Deleted 1970. 350mm approx.	£35	£25
Triang Torpedo Boat.		
Cream and red or blue and cream. Clockwork model of superior quality. Price 3/11d. Issued 1952. Deleted 1970. 290mm.	£15	£10
Triang Gyro Speed Boat.		
White and blue or red and blue. After a few turns of the strong gyro mechanism this boat travels quite a long distance. Very popular, as one did not fear breaking any springs. Price 4/11d. Issued 1950. Deleted 1965. 290mm.	£15	£10
Doll's Pushchair with Hinged Hood.		
Red and cream or blue and cream with strong steel wheels and hinged black hood. Price 36/-. Issued 1953. Deleted 1970. 115mm approx.	£15	£10
Tin-Plate Chicken with Cart and Small Chick.		
A mixture of red, cream, blue and gold. Price 4/11d. Issued 1950. Deleted 1970. 200mm.	£25	£20
Puff-Puff Train with Whistle.		
Blue or red with 'Triang' decals. Price 17/6d. Issued 1951. Deleted 1970. 455mm approx.	£25	£20
Triang Handcart.		
Strong wooden body with red or blue steel wheels, rubber tyres and handle. Complete with a set of nursery bricks. A popular toy. Price 17/6d. Issued 1952. Deleted 1970. 660mm.	£15	£10

MODEL	MB	MU	GC

The Pecking Bird.

Blue, green and white with fine clockwork action. Yellow legs with red tips on claws. Price 2/11d. Issued 1950. Deleted 1970. 120mm.

	£25	£20	£15

Jack in The Boat.

Red boat with brown or fawn interior, cream oars and figure of man in blue. Strong clockwork motor that made the oars move at a high speed through the water. Tinplate and plastic. Price 6/11d. Issued 1952. Deleted 1970. 242mm.

	£50	£45	£35

Barnacle Bill with Polly Parrot.

Sailor in blue uniform and white hat with blue band with 'Minic Triang' on brim. The clockwork figure has a red and white bag in one hand and a red and blue parrot in a gold cage in the other. Price 7/11d. Issued 1955. Deleted 1970. 35 x 180mm.

	£50	£45	£35

Dancing Prince and Princess.

Clockwork figures that dance around, when wound up. In blue and cream. A good investment and one of the best Minic toys made. Price 7/6d. Issued 1956. Deleted 1970. Height 180mm.

	£35	£25	£15

Prince in red and princess in golden dress. Details otherwise as above. Rare colours.

	£50	£45	£40

Loch Ness Monster.

Green with white and yellow stripes, large head and black and white eyes. Clockwork. Price 3/9d. Issued 1954. Deleted 1970. 520mm.

	£20	£15	£10

No.3 Minic Electric Railway.

This tin-plate toy was a popular item and came in green with a miniature clockwork train, two carriages and trucks in brown and yellow. Complete with tunnels, crossings, gates and scenic layouts. 'Triang Minic Railway' in white letters. Price 22/6d. Issued 1953. Deleted 1970. 560 x 109mm

	£40	£35	£25

No.3/A. Railway Set in green with blue and dark maroon train set, otherwise as above.

	£35	£30	£20

The Skyliner Plane.

Red with black line design around windows and along centre. Clockwork. Black wheels and cream interior cockpit. Price 4/6d. Issued 1952. Deleted 1970. 391mm.

	£25	£20	£15

The Sky King Airliner.

Grey with cream tint, red and blue design along centre, silver props with black wheels, otherwise as above.

	£25	£20	£15

MODEL	MB	MU	G
Minic Ladybird.			
Brown and black with fine clockwork action. Price 1/11d. Issued 1953. Deleted 1970. 91mm.	£9	£7	£5
The Magic Puffin.			
White and dark blue or black with orange flipper feet and strong clockwork motion. Ringed nose and red eyes. Price 3/6d. Issued 1953. Deleted 1970. 180mm.	£15	£10	£8
A Minic Land Yacht.			
Red boat with white sail and black wheel. Clockwork. Price 5/11d. Issued 1954. Deleted 1970. 266mm.	£15	£10	£8
Minic Mouse.			
Strong clockwork action. Grey with little pink ears. Price 1/6d. Issued 1954. Deleted 1970. 91mm	£10	£8	£5
Mighty Jabberwock.			
Dark orange with black tail and black and dark brown dots. Strong clockwork action. Price 1/9d. Issued 1954. Deleted 1970. 300mm.	£15	£10	£8
Great Grabbing Spider.			
Medium blue with long brown legs and red shoe-shaped feet. Dark blue or black striped design. Price 4/11d. Issued 1954. Deleted 1970. 120mm.	£15	£10	£8
Minic Nuffield Tractor. Clockwork.			
Grey with brown wheels and pipe. Large black tyres at rear and smaller at front. Driver in blue shirt and black trousers. Price 6/6d. Issued 1950. Deleted 1970. 208mm.	£20	£15	£1
No.2 Nuffield Tractor and Driver. Clockwork.			
Dark orange with orange-and-pink-clad driver. Large wheels with grey tyres at rear and smaller tyres at front. Price 7/11d. Issued 1953. Deleted 1970. 208mm.	£15	£10	£8
No.3 Service Station.			
Dark orange, cream and white with red roofs. Complete with three cars, a van and a breakdown truck, all with strong clockwork motors. A winch lift for taking cars and goods to the roof and top floor. Price 48/11d. Issued 1954. Deleted 1970. 408mm approx.	£30	£25	£2
No.3/A. Minic Service Station.			
Cream, light blue and white, complete with six cars, two vans and a breakdown truck, all clockwork. Two-storey garage with 'Minic' decals. Price 49/11d. for garage only, or 71/- complete with cars etc. Issued 1954. Deleted 1970. 620mm approx.	£35	£30	£2
Minic No.2 'Sherman Tank' Clockwork.			
Olive green or yellow green with revolving turret and black tracks. Twelve roller wheels and firing gun. White star decals. Price 5/11d. Issued 1952. Deleted 1972. 216mm.	£30	£25	£2
No.2 Armoured Car.			
Dark army grey with swivel gun turret and thick wheels with grey tyres. Price 3/11d. Issued 1950. Deleted 1970. 158mm.	£25	£20	£1
No.2/A. Dark grey, brown and black battlefield livery. Price 3/11d. Issued 1950. Deleted 1970. 158mm.	£30	£25	£2

MODEL	MB	MU	GC
Grand Prix Racing Car Clockwork.			
Medium red with large tin-plate wheels, grille and thick grey tyres. Driver in brown suit and dark brown helmet. No.9 on rear and nose. Price 5/11d. Issued 1950. Deleted 1970. 128mm.	£20	£15	£10
No.2 Grand Prix Racing Clockwork Model.			
Medium blue with silver tint, silver grille, wheels with thick black tyres and driver in white. With authentic engine noise from clockwork motor. Price 5/6d. Issued 1952. Deleted 1970. 128mm.	£25	£20	£15
Musical Saloon. Clockwork.			
Dark brown with metal grille, bumpers etc. Price 2/11d. Issued 1949. Deleted 1970. 118mm.	£30	£25	£20
No.2 Remote Control Electric Pathfinder.			
Bright red with grey interior, metal bumpers, grille and wheels. Black tyres with panel-control unit in red. Price 8/11d. Issued 1953. Deleted 1970. 180mm	£25	£20	£15
No.2A Pathfinder.			
Dark brown, otherwise as above.	£30	£25	£20

MODEL	MB	MU	GC
No.2. Taxi.			
Yellow with silver grille, bumpers, wheels and grey tyres, brown interior and roof rack, strong clockwork motor. Price 3/6d. Issued 1950. Deleted 1970. 118mm.	£25	£20	£15
No.2 Musical Saloon.			
Dark blue with brown trim, pink interior with strong clockwork motor that connects to a mechanical musical box which plays as car in motion. Silver bumpers, grille, headlights and wheels with grey tyres. Price 4/11d. Issued 1951. Deleted 1970. 118mm.	£25	£20	£15
London Fire Engine.			
Red with silver grille, bumpers, golden bell, black mudguards, red ladder. Price 4/6d. Issued 1955. Deleted 1970. 182mm.	£18	£15	£10
Clinic Jeep. Clockwork.			
Dark green with strong clockwork motor, metal wheels, bumpers, grille and grey tyres. Spare wheel on rear. Price 2/11d. Issued 1950. Deleted 1970. 182mm.	£12	£10	£8
No.2 Jeep. Clockwork.			
Red or rose pink with pink interior, red grille, bumpers, metal wheels and grey tyres. Spare wheel on rear. Price 3/6d. 182mm.	£10	£8	£6

Minic Car and Caravan.

Car in rich medium blue with silver grille, bumpers, wheels, rust interior, black mudguards, tow-bar and silver headlights. Caravan in red with pinkish cream door, sloping roof and metal wheels with black tyres. Strong clockwork motor. Price 9/11d. Issued 1955. Deleted 1970. 270mm.

| | £30 | £25 | £2 |

Red car with red and white caravan, otherwise as above.

| | £35 | £30 | £2 |

Black car with blue and cream caravan, otherwise as above.

| | £50 | £45 | £4 |

Minic Traction Engine and Trailer.

Matching green with black trim on engine and trailer. Yellow pipe and boiler, spoked rubber wheels. A good forward and backward movement from strong clockwork motor. Price 7/11d. Issued 1955. Deleted 1970. 242mm.

| | £30 | £25 | £2 |

Traction Engine and Tar Barrel.

Red traction engine with black pipe and boiler, gold line trim and spoked wheels with black tyres. Tar barrel trailer in matching red. Clockwork action. Price 7/11d. Made in a limited number in 1955. 222mm.

| | £30 | £25 | £2 |

Minic Jeep. Clockwork.

Military light green with metal wheels and large grey knobby tryes, star decal in white on bonnet. Spare wheel on rear. Price 3/6d. Issued 1957. Deleted 1970. 87mm.

| | £15 | £10 | £7 |

No.57. Farm Tractor. Clockwork.

White metal and silver radiator surround and grille, medium yellow body with matching pipes, grey steering wheel and metal wheels. Large black tyres at rear and smaller ones at front. Knobby treads. Price 4/6d. Issued 1958. Deleted 1970. 87mm.

| | £15 | £10 | £7 |

No.57/A. Red body with driver, othrwise as above.

| | £30 | £25 | £2 |

Minic Racing Car. Clockwork.

Medium sky blue body with red exhausts, driver in orange, silver grille, wheels with large black or grey tyres. Red number '6' in white circle. Price 4/11d. Issued 1956. Deleted 1970. 135mm.

| | £20 | £15 | £1 |

Red body with black exhausts, otherwise as above.

| | £25 | £20 | £1 |

Minic Tractor.

Orange body, seat and steering wheel. Large grey wheels and grey tracks, silver grey grille. Original price 3/-. Issued 1957. Deleted 1970. 87mm.

| | £20 | £15 | £1 |

Tractor with blue body, otherwise as above.

| | £25 | £20 | £1 |

Tractor in red with black tracks and wheels, otherwise as above.

| | £30 | £25 | £2 |

Minic Double Decker Money Box.

The year 1964 was very competitive with gimmicks of all kinds connected with toys. Minic made a unique double decker bus in red with a slot in the roof for coins. Friction driven and plastic, this proved to be a great seller. Various decals came with this model. Price 5/11d. Issued 1964. Deleted 1970. 202mm.

| | £15 | £12 | £1 |

Minic Steamroller. Clockwork.

Green body with grey rollers at front and large grey wheels at rear. Gold boiler and funnels. Price 7/6d. Issued 1959. Deleted 1970. 130mm.

| | £15 | £12 | £9 |

Minic Clockwork Bulldozer.

Rust or dark orange with black tracks and large blade with 'Triang' decal in black on blade. Black engine. Price 5/11d. Issued 1959. Deleted 1970. 120mm.

	£18	£15	£10

Minic Ambulance.

Grey body with silver grille and bumpers. Wheels with grey or black tyres. White crosses on red circles on sides, roof and rear doors, which open. Clockwork motor. Price 4/11d. Issued 1958. Deleted 1970. 128mm.

	£18	£15	£10

Cream body and grey tyres, red crosses on white circles, otherwise as above.

	£20	£18	£15

Minic London Taxi. Clockwork.

Royal blue with smart silver grille, bumpers, headlights, wheels and black tyres. Black running board, meter, open passenger/luggage door. One of the rarest and most sought Minic. Price 7/6d. Issued 1959. Deleted 1970. 108mm.

	£20	£15	£12

Black body with grey or black tyres, otherwise as above.

	£25	£20	£18

Minic Bulldozer.

Dark green with grey blade and 'The British Triang Co.' decals in black. Grey tracks and pipe. Driver in red shirt. Price 5/9d. Issued 1950. Deleted 1970. 216mm.

	£20	£15	£10

No.2. Minic Bulldozer.

Dark orange with black tracks and pipe. Driver with green shirt. Price 7/6d. Issued 1953. Deleted 1970. 216mm.

	£25	£20	£15

Minic Clockwork Reverse Control Steamroller.

Green with large grey wheels at rear and double rollers in grey at front. Golden pipes and boiler, black line trim. Clockwork motion backwards and forwards. Price 7/11d. Issued 1959. Deleted 1970. 100mm.

	£20	£15	£12

With large black wheels and small double rollers, silver boiler and pipes otherwise as above.

	£25	£20	£15

Taxi. Clockwork.

Black with tin wheels, grille, bumpers and windscreen. Price 2/1d. Issued 1949. Deleted 1970. 118mm.

	£20	£15	£10

Minic Tank.

Dark green with white star decals. With white tracks. Price 3/6d. Issued 1950. Deleted 1970. 158mm.

	£15	£12	£9

No.4. Service Station.

Stone, white and orange with red-tiled roof. Complete with 'Minic Service Station' decals and signs, six petrol pumps and one oil pump. The upper storey can be used as a showroom for Minic cars. Roof lifts off and the glass window has a metal frame. Equipment includes petrol pumps, oil cabinets, roof sign, car joist, turntable, car and road ramps. Price 50/-. Issued 1947. Deleted 1970. 61cms.

£40 £35 £

No.10. Minic Garage.

Stone, white, red and black. A realistic model fitted with sliding shutter, complete with three petrol pumps and clock above garage door. Price 12/6d. Issued 1947. Deleted 1970. 22cms approx.

£20 £15 £

No.77M. Double Deck Trolley Bus.

Red and cream with 'Bovril' and 'London Transport' decals. One of the rarer models that everyone dreams of finding. Price 3/6d. Issued 1948. Deleted 1970. 180mm.

£50 £45 £

Green and cream, with 'Triang Minic' advert, otherwise as above.

£60 £55 £

POST-WAR COMMERCIALS

MODEL	MB	MU	GC

6 Ton Tipper. Clockwork

Green cab with white interior and yellow tipping body. Metal
wheels, grille and bumpers. Black chassis and large grey tyres. Price
6/11d. Issued 1950. Deleted 1970. 195mm.

£25 £20 £15

No.2. 6 Ton Tipper. Clockwork.

Green with white cab interior, red tipping body, silver bumpers,
grille etc. Price 7/6d. Issued 1954. Deleted 1970. 195mm.

£20 £15 £10

Double Decker London Bus.

In red and light pink or cream with 'Minic Cars' decals, metal
grille, bumpers, wheels and headlights. Black bumpers on some
models with black mudguards. Grey tyres. Strong clockwork motor.
Price 12/6d. Issued 1955. Deleted 1970. 512mm.

£25 £20 £15

Minic London Single Decker Bus.

Green body with yellow line trim. 'Green Line' decals in yellow,
silver grille, bumpers, wheels black tyres and neat mudguards.
Brown seats inside. Clockwork. Price 6/9d. Issued 1955. Deleted
1970. 512mm.

£25 £20 £15

Red body with black line trim. 'Midland Coaches' decals, otherwise
as above.

£40 £35 £25

Mechanical Horse and Trailer with Six Foot Cruiser.

Blue cab with dark blue interior, lighter blue trailer, metal wheels,
grille, bumpers and black tyres. Red and fawn boat. Both vehicles
have strong clockwork motors. Price 17/11d. Issued 1955. Deleted
1970. 216mm.

£25 £20 £15

Minic Breakdown Truck. Clockwork.

Green with red cab and brown interior. Red crane, white hook,
silver grille, bumpers. Price 4/11d. Issued 1955. Deleted 1970.
121mm.

£18 £15 £10

Mechanical Horse and Cable Drum Trailer.

Red cab with dark red interior, silver metal grille, bumpers and
wheels. Medium blue trailer with two cable drums in red with grey
surrounds. All metal wheels and thick rubber tyres. Clockwork.
Price 8/11d. Issued 1955. Deleted 1970. 218mm.

£20 £15 £10

Blue cab with dark blue interior, otherwise as above.

£30 £25 £20

Morris Light Van.

Medium blue with lighter blue interior, metal bumpers, wheels and
rubber tyres. Clockwork. Decals on sides. Price 2/11d. Issued 1956.
Deleted 1970. 91mm.

£15 £12 £9

Clockwork Horse and Watney Trailer.

Green cab with dark brown interior, grey or mid-blue grille,
bumpers, wheels and black mudguards. Small pink barrel on roof of
cab with 'Watney's' sign. Large barrel in fawn with black hoops
and 'Watney' decals. Price 7/6d. Issued 1956. Deleted 1970.
175mm.

£20 £15 £10

Minic Lorry. Clockwork.

Red cab with dark red interior, green chassis and wagon body, tin
grille, bumpers, wheels with grey tyres. Price 5/11d. Issued 1957.
Deleted 1970. 140mm.

£18 £15 £10

British Road Services Lorry. Clockwork.

In dark maroon with metal bumpers, grille, headlights, wheels and black tyres. Price 6/11d. Issued 1957. Deleted 1970. 135mm.   £18   £15   £1(

Austin A40 Van. Clockwork.

Light green with tin grille, bumpers, wheels, and black tyres. 'Minic Motors' decals on sides in red and white. Price 3/11d. Issued 1957. Deleted 1970. 87mm.   £12   £10   £7

Morris 10cwt. Van. Clockwork.

Maroon with 'Triang Minic' decals in red and gold on sides. Silver grille, wheels and neat black tyres. Price 5/11d. Issued 1957. Deleted 1970. 91mm.   £14   £12   £9

Royal Mail Van.

Post office red with crown and 'Royal Mail' emblems etc., metal grille, mudguards, bumpers, wheels and black tyres. Clockwork. Price 2/11d. Issued 1956. Deleted 1970. 91mm.   £15   £10   £8

Morris Telephone Van. Clockwork.

In telephone post office green with decals, crown etc., metal bumpers, red ladder on roof, metal grille, wheels and grey tyres. Price 2/11d. Issued 1956. Deleted 1970. 91mm.   £15   £10   £8

Minic Timber Lorry. Clockwork.

Blue cab with dark interior, silver grille, bumpers and headlights, red mudguards, chassis and body with timber load. Metal wheels and black tyres. Price 6/6d. Issued 1957. Deleted 1970. 136mm.   £16   £14   £12

Minic Dump Truck. Clockwork.

Golden yellow body with metal wheels, large at rear, small at front, thick knobby tyres. Price 4/6d. Issued 1957. Deleted 1970. 102mm.   £14   £12   £9

Clockwork Tipper Lorry.

Red cab with dark red or brown interior, black chassis and medium blue tipper with 'Triang Minic Services' decals in red and white on sides. Winch action with metal bumpers, grille and wheels with grey tyres. Price 5/11d. Issued 1957. Deleted 1970. 114mm.   £16   £14   £11

Minic Shutter Van. Clockwork.

Red body with 'Minic Transport', 'Royal Mail', and 'Express Service' decals in red on cream oblong design on sides. Shutters at rear. Silver grille, bumpers and wheels with grey or black tyres. Price 4/11d. Issued 1958. Deleted 1970. 128mm.   £18   £15   £12

Minic British Road Services Van.
With an extra strong clockwork motor, this van is in green with
British Road Services' decals in white along top sides of model.
Metal grille, headlights, bumpers and wheels with black tyres. Price
4/11d. Issued 1958. Deleted 1970. 135mm.

	£18	£15	£12

Minic Tractor and Trailer with Cases.
Tractor in dark lemon or mustard with silver grille and steering
wheel. Large wheels and grey tyres on rear and smaller ones on
front. Green trailer with yellow lines and cream box load. In special
gift box. Price 10/11d. Issued 1954. Deleted 1970. 182mm.

	£20	£18	£15

Minic Mechanical Horse & Log Trailer.
Red cab with dark interior, silver grille, bumpers and headlights
and wheels with grey or black tyres. Log trailer in black with two
large logs with chains. Clockwork. Price 10/6d. Issued 1954.
Deleted 1970. 191mm.

	£18	£15	£12

Mechanical Horse & Horsebox.
Cream cab with red interior, maroon horse-box with opening doors
at side and rear. Spare tyre and wheel and two horses. Headlights,
bumpers, grille and matching silver wheels. 'Minic Horse
Transport' decals on sides in red and black. Very rare. Price 12/6d.
Issued 1954. Deleted 1970. 206mm.

	£20	£18	£15

Minic Heavy Transport Van. Clockwork.
Purple cab with dark red interior, light blue body with 'Minic
Transport, Road, Rail, Air and Sea Express Services' decals in red
and black on cream background. Silver grille, bumpers, headlights
and wheels and black tyres. Price 6/11d. Issued 1955. Deleted 1970.
135mm.

	£20	£18	£15

Cream cab and red box body, otherwise as above.

	£25	£20	£18

Minic Clockwork Petrol Tanker.
Green cab with dark interior, red tanker body in Shell livery and
'Shell' decals in very large golden yellow letters. Silver grille,
bumpers and wheels and grey tyres. Price 7/11d. Issued 1955.
Deleted 1970. 155mm.

	£20	£18	£15

Minic Clockwork Tanker.
Dark blue cab with green tanker body. 'British Petroleum' decals in
black. Metal grille, bumpers, wheels and thick grey knobby tyres.
Price 4/11d. Issued 1955. Deleted 1970. 155mm.

	£20	£18	£15

Clockwork Case Lorry.
Green cab with red interior, tin-plate grille, bumpers, headlights
and wheels with black or grey tyres with knobby treads. Red body
with six cream or white packing cases. Price 5/6d. Issued 1955.
Deleted 1970. 150mm.

	£15	£12	£10

MODEL	MB	MU	G

Minic Mechanical Horse & Super Pantechnicon.

Red horse with dark brown interior and large purple box body with
'Minic, Road, Rail, Air & Sea' decals in red, black and cream.
Silver grille, bumpers, headlights and wheels with large black or
grey knobby treads. Price 8/11d. Issued 1956. Deleted 1970.
182mm. £18 £15 £1

Large Mechanical Horse & Fuel Tank Trailer.

Green horse body with black interior, silver grille, headlights,
bumpers and wheels with large black tyres. Red tanker body with
gold lettering. 'Shell B.P. Fuel Oil' decals. Price 9/11d. Issued
1956. Deleted 1970. 182mm. £18 £15 £1

Red body on horse with green tanker body, otherwise as above. £25 £20 £1

Ice Cream Van. Clockwork.

Cream and blue two-tone with figure serving cone. Silver grille,
wheels, bumpers and opening doors. Opening hatch at side. Price
5/11d. Issued 1959. Deleted 1970. 108mm. £18 £15 £1

Clockwork Bakery Van.

Dark brown body with red interior, metal grille, headlights,
bumpers and wheels with black or grey tyres. Oval shaped windows
on rear and sides of cab. Driver in brown or cream with white cap.
'Minic Bakery. The sign of Good Bread' decals in gold lettering.
Price 8/11d. Issued 1959. Deleted 1970. 128mm. £18 £15 £1

Mechanical Horse & Trailer with Cases.

Large red cab with dark interior and large green wagon body with
six large box-type packing cases in cream or coffee. Silver grille,
headlights, bumpers, wheels and black tyres. Clockwork motor.
Price 7/6d. Issued 1953. Deleted 1970. 182mm. £18 £15 £1

Clockwork Old London Taxi.

Purple body with black roof and mudguards. Silver bumpers,
headlights, grille and wheels and grey tyres. Price 7/11d. Issued
1953. Deleted 1970. 108mm. £20 £18 £1

Green body with black running-board and mudguards, otherwise as
above. £30 £25 £2

Minic Luton Clockwork Van.

Red and dark purple with dark red cab interior. Metal grille,
wheels, bumpers and black wheels. 'Minic Transport. Express
Service' decals in red and black on sides. Price 5/11d. Issued 1956.
Deleted 1970. 180mm. £15 £12 £9

British Railways Van. Clockwork.

Cream upper body, red lower body, dark cab interior, silver grille,
bumpers, headlights and wheels with black tyres. 'British Rail'
decals in red, black line design around centre of van with opening
rear doors. Price 7/11d. Issued 1956. Deleted 1970. 137mm. £20 £18 £1

Clockwork Minic Refuse Lorry.

One of the best tin-plate models of this period. Red and green with
polished tin shutter lids on both sides of wagon body. Headlights,
grille, bumpers and wheels with black tyres. 'Triang' decals in
black letters on red square. Price 6/6d. Issued 1956. Deleted 1970.
180mm. £18 £15 £1

MODEL	MB	MU	GC
Mechanical Horse and Brewers Trailer. Complete with Barrels.			
Large red cab with dark brown interior. Green wagon with yellow lines and containing 8 fawn wooden barrels. 'Minic Brewery' decals in red, black and yellow on sides of wagon. Tin wheels, grille, bumpers, headlights, and grey tyres. Price 7/6d. Issued 1956. Deleted 1970. 136mm.	£25	£20	£15
Minic Mechanical Horse and Milk Tank Trailer.			
Another popular model with strong clockwork action. Red with white tanker body, with 'Minic Dairies 3,150 Gallons' decals in blue. Silver grille, bumpers, headlights and wheels with large black tyres. Price 9/6d. Issued 1955. Deleted 1970. 202mm.	£20	£18	£15
Post Office Clockwork Van.			
Made of good quality plastic and one of the first of this new material ever produced. In red with black or gold lettering with crown on sides. Price 2/11d. Issued 1960. Deleted 1970. 79mm.	£9	£7	£5
Post Office Telephone Van. Plastic.			
In Post Office telephone green with crown and decals. Clockwork. Price 2/11d. Issued 1960. Deleted 1970. 79mm.	£9	£7	£5
Minic Removal Van. Clockwork and Plastic.			
Dark blue body with white lettering. Price 6/11d. Issued 1961. Deleted 1970. 114mm.	£10	£8	£6
Morris Minor Plastic. Friction.			
Blue body with jewelled headlights, black tyres and windscreen wipers. Price 7/6d. Issued 1961. Deleted 1970. 108mm.	£9	£7	£5
Austin Taxi, Plastic Friction.			
Navy blue or black body with silver grille, headlights and bumpers. Dark interior and high quality outer finish. Price 7/6d. Issued 1962. Deleted 1970. 108mm.	£10	£8	£6
Also with green body, otherwise as above.	£15	£12	£10
Box Van, Plastic Friction.			
Red body and cab with off-white interior, black grille, silver headlights and bumpers. Price 4/11d. Issued 1963. Deleted 1970. 122mm.	£10	£8	£6
Open Lorry in Plastic Friction.			
Cream cab and red body, black grille, bumpers and wheels. Price 5/11d. Issued 1964. Deleted 1970. 136mm.	£9	£7	£5
Double Decker Bus. (Routemaster).			
Red and cream. Also in Blue and Cream, worth treble. This model was the nearest thing to a promotional bus for the London Transport Company. Photographs were supplied to Lines Brothers by London Transport. Please note there are several adverts connected with this model. Some collectors will pay high prices for special adverts.			
'Triang Transport.'	£18	£15	£12
'Try Out Your New London Bus with Routemaster.'	£30	£25	£20
With the rare 'Ribble Motor Co.' in blue and cream.	£50	£45	£35
'Pedigree prams' advert or 'Triang Pedal Motors.'	£20	£15	£10

MODEL	MB	MU	GC

No.2 Sports Car with Horn

Lime green with cream seats and trim. Metal windscreen, bumpers,
wheels and grille. Grey tyres and automatic horn that works in
conjunction with clockwork motor. Price 2/9d. Issued 1950. Deleted

1970. 170mm.	£15	£12	£9
No.2/A Dark red, otherwise as above.	£20	£15	£12
No.2/B Yellow and cream, otherwise as above.	£25	£20	£15

Ford Zephyr Saloon, Clockwork.

Dark orange, red with tin plate grille, bumpers and wheels, dark
interior and black tyres. Price 3/6d. Issued 1950. Deleted 1970.

180mm.	£15	£12	£9
Dark grey, otherwise as above.	£20	£15	£12
Dark green, otherwise as above.	£25	£20	£15

No.1 Sports Car.

Ruby red with cream trim, all metal wheels, bumpers, grille. Strong
clockwork motor and grey tyres. Price 2/-. Issued 1949. Deleted

1970. 155mm.	£20	£18	£15

No.2 Sports Car.

Open sports car with four speed gearbox, strong clockwork engine,
medium blue body, red seats with silver trim lines, bumpers, grille,
headlights and metal wheels with grey tyres. Price 2/11d. Issued

1951. Deleted 1970. 155mm.	£15	£12	£9

Minic Pathfinder Electric Car.

The first electric remote control model made by Triang Minic.
Golden yellow with black line trim, metal wheels, bumpers and
grille. Separate control panel with steering wheels which enables
one to steer the model backwards or forwards. Price 7/11d. Issued

1950. Deleted 1970. 180mm.	£20	£15	£12

No.2 Ford Zephyr Saloon. Clockwork.

Blue with dark purple tint, dark brown interior, silver grille,
bumpers, wheels etc. Price 3/11d. Issued 1953. Deleted 1970.

180mm.	£15	£12	£8

Vanguard Saloon. Clockwork.

Dark blue with tin plate grille, bumpers, wheels with grey tyres.

Cream interior. Price 3/11d. Issued 1950. Deleted 1970. 180mm.	£14	£12	£8
Deep purple, otherwise as above.	£20	£15	£10

MODEL	MB	MU	GC

NO.2 Vanguard Saloon. Clockwork.

Green with golden lines, cream interior, solid wheels and thick grey tyres. Silver bumpers, grille etc. Price 5/6d. Issued 1953. Deleted 1970. 191mm.   £15   £12   £9

Stop-on Saloon. Clockwork.

Bright green with golden trim and silver grille, bumpers and wheels. Grey tyres, opening boot, brown interior. Price 2/6d. Issued 1950. Deleted 1970. 158mm.   £14   £12   £10

No.2 Stop-on Saloon. Clockwork.

Rose-red with brown interior, silver grille, bumpers etc, and grey tyres. Price 3/6d. Issued 1951. Deleted 1970. 180mm.   £12   £10   £8

Police Car. Clockwork.

Dark blue with roof sign and aerial. Tin radiator, bumpers, wheels etc. Pink interior. Price 4/6d. Issued 1949. Deleted 1970. 180mm.   £18   £15   £12

NO.2 Police Car. Clockwork.

Dark maroon with roof sign, silver grille, bumpers, wheels with grey tyres. Light pink interior. Price 3/11d. Issued 1951. Deleted 1970. 180mm.   £15   £12   £9

Minic Buick Sedan.

Red or rose pink with dark brown interior and silver grille and bumpers. Black wheels and tyres. Price 4/11d. Issued 1954. Deleted 1970. 160mm.   £15   £12   £9

Minic Control Special Clockwork Car.

Silver body with dark blue running board, silver grille, bumpers, wheels with grey tyres. Driver in black uniform, lady and child passenger. Rare model. Price 8/11d. Issued 1955. Deleted 1970. 180mm.   £20   £15   £10

Rolls Sunshine Saloon. Clockwork.

Ruby red with black mudguards and and sliding sunshine roof. Silver grille, headlights, bumpers, cream interior, metal wheels and grey tyres. Price 6/11d. Issued 1954. Deleted 1970. 160mm.   £15   £12   £9

Green body, otherwise as above.   £20   £15   £10

Bright blue body and blue interior, otherwise as above.   £26   £20   £15

Jowett Javelin Car.

Pale blue with black tyres, silver grille, bumpers etc and white interior. Price 2/11d. Issued 1955. Deleted 1970. 121mm.   £15   £12   £9

Riley Saloon.

Medium purple with black interior, silver grille, bumpers, headlights and wheels. Purple mudguards, black tyres. Strong clockwork motor. Price 3/11d. Issued 1955. Deleted 1970. 121mm.   £15   £12   £9

Ford Monarch Sedan Special.

Lime green with strong clockwork motor, metal bumpers and wheels and black tyres. Black interior. Price 3/11d. Issued 1955. Deleted 1970. 121mm.   £14   £11   £8

Red, otherwise as above.   £20   £15   £10

Minic Rolls Sedan Car.

This fine clockwork open tourer was a best seller when introduced. Light green with gold tint, metal grille, bumpers, wheels and black mudguards. Black rear half canopy, white seats, black steering wheel. Opening door. Price 5/6d. Issued 1955. Deleted 1970. 146mm.

	£18	£15	£1

Red body with other details as above apart from pink seats.

	£25	£20	£1

Minic Roomy Sports Special.

Grey body with all silver mudguards and black number '7' on bonnet and doors. Spare wheel at rear. Strong clockwork motor. Price 6/-. Issued 1955. Deleted 1970. 166mm.

	£18	£15	£1

Minic Rolls Tourer.

Silver blue with black mudguards, silver bumpers and grille, blue seats and grey tyres. Open tourer with strong clockwork motor. Price 3/9d. Issued 1955. Deleted 1970. 146mm.

	£15	£12	£9

Red body with black tyres, otherwise as above.

	£20	£15	£1

Minic Morris Minor.

Red body, metal bumpers, grille, etc. and dark interior. Issued 1956. Deleted 1970. Price 2/6d. 90mm.

	£12	£10	£7

The Minic 'O' Saloon

Rose red body with black interior, metal grille, bumpers and wheels. Rubber tyres. Clockwork. Price 3/6d. Issued 1956. Deleted 1970. 67mm.

	£14	£12	£1

Minic Morris Minor. Clockwork.

Blue with dark brown interior, metal grille, wheels and bumpers and grey tyres. Price 2/6d. Issued 1956. Deleted 1970. 90mm.

	£14	£12	£1

Hillman Minx Saloon. Clockwork.

Maroon with dark maroon interior, metal grille, bumbers and wheels and grey tyres. Price 2/11d. Issued 1956. Deleted 1970. 91mm

	£15	£12	£1

Morris Oxford Saloon. Clockwork.

Purple body with dark interior, metal grille, bumbers and wheels and black tyres. Price 3/6d. Issued 1956. Deleted 1970. 102mm.

	£14	£12	£9

In red, otherwise as above.

	£20	£15	£1

Standard Vanguard. Clockwork.

Green with darker interior, metal grille, bumpers and wheels and black tyres. Price 3/6d. Issued 1956. Deleted 1970. 102mm.

	£15	£12	£1

In black, otherwise as above.

	£20	£15	£1

In dark grey, otherwise as above.

	£25	£20	£1

MODEL	MB	MU	GC

Minic Austin A40 Saloon. Clockwork.

Lime green with yellow tint, silver grille, bumpers and wheels and grey tyres. Black interior. Price 3/11d. Issued 1957. Deleted 1970. 91mm.

	£15	£12	£9

Fawn two tone body, otherwise as above.

	£20	£15	£12

Minic Streamline Sports. Clockwork.

Medium green body with brown or dark red seats, open tourer with silver grille, bumbers and wheels, grey tyres. Price 3/6d. Issued 1956. Deleted 1970. 128mm.

	£20	£18	£15

Minic Clockwork Hurricane Armstrong Siddeley.

Deep purple body with dark interior, silver grille, headlights and bumpers, wheels with black tyres. Price 4/11d. Issued 1957. Deleted 1970. 136mm.

	£15	£12	£9

Ruby red body and grey tyres, otherwise as above.

	£25	£20	£15

Traffic Control Car. Clockwork.

Light blue with grey or blue interior, silver grille, headlights and bumpers and wheels with grey or black tyres. Black running boards and roof lights. Price 6/11d. Issued 1959. Deleted 1970. 128mm.

	£18	£15	£12

Standard Vanguard. Plastic Friction.

Medium grey with silver grille, headlights, bumpers and wheels. Pric 7/11d. Issued 1963. Deleted 1970. 108mm.

	£8	£6	£4

Rover 90. Friction Plastic.

Dark fawn with silver grille, headlights, bumpers and wheels. Price 8/11d. Issued 1964. Deleted 1970. 108mm.

	£7	£5	£3

SPOT-ON VEHICLES

MODEL	MB	MU	G

No.100. Ford Zodiac

Two tone blue with off white or yellow mustard interior, silver bumpers, grille, wheels and trim. This was the pride of the Ford range, always renowned for its vivid and quick acceleration at high cruising speed. This was the first model ever produced by the Spot On Company. Price 4/11d. Issued 1959. Deleted 1966. 111mm.

£15 £10 £5

Also in blue and fawn two tone, otherwise as above. £12 £9 £7

No.100 S/L Ford Zodiac.

Red and fawn with golden interior, silver wheels, grille, bumpers and lights. Price 6/6d. New design issued 1963. Deleted 1966. 111mm.

£20 £15 £1

No.101 Armstrong Siddeley Sapphire 236.

Metallic red with black roof and dark rose interior. A specially built model, like the car itself, with a no clutch system that actually thought for the driver. Price 4/11d. Issued 1959. Deleted 1966. 108mm.

£30 £25 £1

Also in golden yellow with black roof and full silver trim, otherwise as above. £15 £10 £7

No.102 Bentley Four Door Sports Saloon.

Two tone red upper half and silver grey lower half, orange interior and silver trim, headlights, grille etc. This was the most luxurious sports saloon of its day, developed from a long line of racing cars. Price 5/-. Issued 1959. Deleted 1965. 127mm.

£20 £15 £1

No.103 Rolls Royce Silver Wraith

Two tone, silver grey or blue grey on the upper and metallic bronze on lower. Still regarded as the world's finest car. Price 5/6d. Issued 1959. Deleted 1962. 127mm.

£30 £25 £1

No.103 Rolls Royce Silver Wraith. Second Issue.

Two tone silver and deep maroon. New design for 1961. This was the model which oil kings and royal families have owned ever since its first appearance. It has won more awards than all the other makes of cars put together. It had a six cylinder 4887 cc engine, overhead inlet and side exhaust valves. Faster than 140 mph — passengers inside hardly knew that the car was moving. The perfect touring limousine. Price 5/6d. Issued 1961. Deleted 1966. 127mm.

£20 £15 £1

No.104 MGA Sports Car.

Red with mustard seats and full silver trim. The favourite of the enthusiast who drives hard and well. Its 4-cylinder 1,489 cc engine develops 72 b.h.p. at 5,500 r.p.m. and in 1956 the MGA took many class 'F' records in the U.S.A. An especially modified car maintaining an average speed of 141.71 m.p.h. for 12 hours. Price 4/6d. Issued 1959. Deleted 1966. 96mm.

£15 £10 £7

MODEL	MB	MU	GC

No.104 MGA Sports Car. Second Issue.
Golden yellow or mustard with red interior and full silver trim,
otherwise as above. Price 5/-. Issued 1960. Deleted 1963. 96mm. £30 £25 £20

No.105 Austin Healey 100-6.
Green body with mustard or dark yellow seats, silver trim,
headlights etc. This was classed as the occasional 4-seater sports
tourer. It was powered by a 6 cylinder 2,639 cc engine, giving 102
b.h.p. at 4,600 r.p.m. This car won many top awards overseas.
Price 4/6d. Issued 1960. Deleted 1965. 86mm. £15 £10 £7

No.106 The Baby Austin.
Black and one of the rarest in the car series. It became almost
unobtainable as soon as it reached the shops. Also in maroon,
which is very rare. Price 3/11d. Issued 1962. Deleted 1965. 54mm. £40 £35 £25

No.107 Jaguar XK SS.
Red body with silver trim and yellow interior. One of the fastest
sports cars ever made. A modified version of the 'D' type, which
was so successful at Le-Mans. Fitted with disc brakes on all wheels.
Could go more than 100 m.p.h. Price 4/11d. Issued 1959. Deleted
1966. 91mm. £12 £9 £7

No.108 Triumph TR3.
Green body, silver trim, orange seats with silver windscreen, wheels
and radiator. Price 4/6d. Issued 1960. Deleted 1966. 89mm. £15 £10 £7

No.108 Triumph TR3, Second Issue.
Silver body with yellow or gold seats and gold or silver wheels.
Rare livery. Price 5/-. Issued 1963. Deleted 1966. 89mm. £25 £20 £15

No.110 Ford Popular.
Black and classed as the working man's car. Reliable, cheap to run
and if one ran out of petrol it free-wheeled for many miles. Price
4/11d. Issued 1963. Deleted 1966. 89mm. £25 £20 £15

No.111 Morris 8 Tourer.
Red or green. Also blue with black seats. This was the great rival
to the Ford Popular car. Price 4/11d. Issued 1960. Deleted 1965.
89mm. £15 £10 £7

No.112 Jensen 541.
Yellow body with black roof and silver trim. This individually built
car with a glass fibre roof was another best seller in its day.
Maximum speed 125 m.p.h., servo-assisted brakes, 4,000 cc engine.
It had the reputation of the car that never lost its value. Price 5/6d.
Issued 1959. Deleted 1966. 108mm. £15 £12 £10

No.113 Aston Martin DB3.
Dark green with orange interior and full silver trim, headlights,
grille etc. Another model developed from a long line of racing cars.
With a tubular classis, an alloy body and disc brakes. Price 5/-.
Issued 1960. Deleted 1966. 105mm. £20 £15 £10

No.114 Jaguar 3.4 litre.
Maroon body with off-white or brown interior. Another sports
saloon and one of the safest fast cars on the road. Price 5/-. Issued
1960. Deleted 1966. 108mm. £30 £25 £20
Green body, otherwise as above. £25 £20 £15

No.115 Bristol 406.

Pale or lime green body with silver trim and lemon interior. A high quality sports car. Price 5/6d. Issued 1961. Deleted 1966. 115mm.

£25 £20 £1

No.115 Bristol 406. Second Issue.

White with matching wheels and dark red interior. Rare colour. Price 5/11d. Issued 1962. Deleted 1966. 115mm.

£45 £40 £30

No.117 Sunbeam motor cycle and sidecar.

In black, silver and gold. A beautiful model made with silver trim, headlights and exhausts. Price 3/11d. Issued 1960. Deleted 1964. 102mm.

£30 £25 £20

No.118 BMW Isetta.

This was a popular bubble car in cream or bright yellow. With number plate, silver wheels and matching trim. Price 3/6d. Issued 1961. Deleted 1966. 56mm.

£15 £10 £7

Also in red, with silver trim, or in green with yellow or off-white interior, otherwise as above.

£25 £20 £15

No.119 Meadows Frisky Sports.

Red with white or blue interior. Made and styled by Italy's great 'Michelotti'. One of the first mini cars in plastic combined with die-cast. A car which did 60 miles to the gallon. Price 2/11d. Issued 1962. Deleted 1966. 69mm.

£12 £10 £7

No.119 Meadows Frisky Sports. Second Issue.

Blue with cream or white interior and full silver trim. Price 3/3d. New design for 1963. Deleted 1966. 69mm.

£15 £12 £10

No.120 Fiat Multipla.

Green with brown or gold interior, silver trim, headlights etc. This was designed from an Italian six seater. One of the most unusual in the small car range. Exceptionally roomy and quite confortable, easy to park and economical to run. Price 2/11d. Issued 1959. Deleted 1966. 91mm.

£12 £9 £7

No.125 Singer Gazelle Saloon.

Green and black, also in red two-tone, full silver trim, headlights, grille etc. Price 5/6d. Issued 1962. Deleted 1966. 95mm.

£25 £20 £15

No.131 Goggomobil (Super)

Pink or rose red body with fawn or golden yellow roof with matching interior, silver trim, headlights, grille etc. This was a German designed and produced coupe, a full-size car with much comfort. It has a high cruising speed with minimum consumption of fuel and many novel technical features including an electro magnetic gearbox and a unique rear engine suspension unit. Price 2/11d. Issued 1960. Deleted 1966. 69mm.

£10 £7 £5

No.133 Utility Jeep.

Dark olive green with lime or off-white interior, silver grey headlights, grille, bumpers etc. Price 2/6d. Issued 1964. Deleted 1966. 69mm.

£15 £12 £9

No.154 Austin A40

Red with a black roof, silver trim, radiator, headlights, grille etc. This was a winner from the Austin stables with its Farina styled body and well tested A35 engine. A firm family favourite that cost little to run. It was reasonable priced when new and is now quite rare, like the Spot-On model itself, which was made in 1960 and deleted in 1966. Price 5/11d. 98mm.

£30 £25 £15

Also in green with black roof. Rare livery, otherwise as above.

£50 £45 £30

No.155 Austin Taxi.

In black with golden interior, full silver trim, grille, headlights etc. With red taxi roof sign. This was a handsome vehicle seen in ever increasing numbers when it appeared on the road and is still in service today in many parts of the country. It replaced the older London cab. It was a luxurious, roomy five seater with a nicely trimmed body and reliable engine. Many were fitted with a 2.2 litre diesel engine with automatic transmission. Price 5/11d. Issued 1961. Deleted 1966. 110mm.

£20 £15 £10

No.157 Rover 3 litre.

Light green with orange or dark green interior. British engineering at its very best. Luxuriously built, finished with a gold trim, headlights, grille and bumpers. Price 7/-. Issued 1960. Deleted 1966. 109mm.

£20 £15 £10

No.158 Rover 3 litre.

Lime green with golden trim and red or orange interior, silver wheels, radiator and bumpers. When first introduced, this Rover was an entirely new body and engine from one of England's leading manufacturers. It represented British engineering skill at its best. Price 5/-. Issued 1960. Deleted 1966. 108mm.

£30 £25 £20

Also in metallic gold, otherwise as above.

£25 £20 £15

Also in red, with matching red interior. A rare colour.

£20 £15 £10

No.159 Morris Minor.

Blue or red body, the latter worth double, with silver trim, headlights and bumpers Price 3/11d. Issued 1963. Deleted 1966. 100mm.

£20 £15 £10

No.165 Vauxhall Cresta.

This was the pride of the Vauxhall range with transatlantic styling. The Cresta represented a satisfying and harmonious blend of the best British and American automobile practice. In royal maroon with golden brown interior, silver line body finish, headlights, grille etc. Price 5/6d. Issued 1960. Deleted 1966. 110mm.

£25 £20 £15

Also in metallic grey with silver trim, otherwise as above.

£35 £30 £20

MODEL	MB	MU	GO

No.166 Renault Floride.

Developed from the famous Dauphine but fitted with a more
powerful engine. The new Floride became a firmer favourite than
its predecessor. Red or rose pink body with full silver trim. Price
6/6d. Issued 1959. Deleted 1965. 102mm.

£25 £20 £15

No.166 Renault Floride. Second Issue.

Metallic gold with orange interior and a black canvas opening top
with full silver trim. Price 6/11d. Issued 1960. Deleted 1966.
102mm.

£30 £25 £20

No.168 Vauxhall Victor Saloon

Another very popular and economical car and always the reliable
engine would outlast the body. The first model was in blue with
silver trim. Price 4/11d. Issued 1963. Deleted 1966. 91mm.

£25 £20 £15

No.183 Humber Super Snipe Estate.

Fawn with white or blue interior, driver, passenger, full silver trim,
roof-rack and luggage. Price 5/11d. Issued 1963. Deleted 1966.
112mm.

£15 £10 £8

No.183 Humber Super Snipe Estate. Second Issue.

Deep or medium red with driver and passenger, roof-rack, luggage
and full silver trim. Price 6/2d. Issued 1964. Deleted 1966. 112mm.

£10 £7 £5

No.184 Austin A60.

Rose red body with full silver trim, headlights, grille and hubs.
With roof-rack and skis. Price 5/6d. Issued 1963. Deleted 1966.
106mm.

£15 £12 £10

Also in green and black two-tone, otherwise as above.

£30 £25 £20

No.185 Humber Saloon.

Green body with driver and full silver trim, headlights, grille etc.
Price 5/-. Issued 1964. Deleted 1966. 112mm.

£15 £12 £9

No.186 Austin Estate Car.

Yellow and dark oak, complete with driver and two passengers.
Price 6/6d. Issued 1965. Deleted 1966. 112mm.

£20 £15 £10

No.191 Sunbeam Alpine.

An entirely new successor to the world famous Sunbeam line. It
became a firm favourite, especially in the U.S.A. Dark fawn with a
black roof, red interior and full silver trim. Price 3/11d. Issued
1962. Deleted 1966. 95mm.

£12 £9 £7

Also in red or rose pink with grey top, otherwise as above.

£9 £7 £5

No.192 Sunbeam Alpine Sports.

A special one-off model made as a rally car with silver metallic
body, complete with driver and flags and designs which varied,
such as Shell, B.P. etc. It had a jet black hard-top with 'No.5' on
doors and nose. Price 5/11d. Issued 1964. Deleted 1966. 95mm.

£20 £15 £10

No.193 NSU Prinz.

White or cream with white interior and full silver trim, headlights,
grille etc. Price 3/11d. Issued 1962. Deleted 1966. 82mm.

£12 £9 £7

Also in alpine red with gold top, otherwise as above.

£15 £12 £10

No.210 Morris Mini Minor Van.

Dark or medium fawn with a green tint, silver wheels, grille, headlights and bumpers. Another winner from the BMC stable. It was roomy and economical, doing 45 m.p.g. It had a front wheel drive, transverse engine and a host of ingenious design features. Price 5/11d. Issued 1960. Deleted 1966. 95mm.

	MB	MU	GC
	£15	£12	£9
Also in lime green. A rare colour, otherwise as above.	£20	£15	£10

No.211 Austin Baby Seven.

A companion to the Mini Minor and in all respects technically identical, although radiator grille differed. Dark or medium blue with yellow or golden interior, silver trim, headlights, grille etc. Price 3/11d. Issued 1960. Deleted 1966. 73mm.

	MB	MU	GC
	£15	£10	£8

No.213 Ford Anglia.

This model was easily recognizable because of the cut back rear window. Two-tone blue with matching interior and full silver trim. Price 3/11d. Issued 1960. Deleted 1965. 95mm.

	MB	MU	GC
	£15	£12	£9
Also in black with off-white interior, otherwise as above.	£25	£20	£15

No.215 Daimler SP 250.

A revolutionary sports car made by one of our oldest manufacturers. Fibre glass body, a V8 engine and disc brakes on all wheels, with a top speed of 120 m.p.h. Red or rose pink with orange or gold seats and full matching silver trim. Price 5/6d. Issued 1960. Deleted 1966. 108mm.

	MB	MU	GC
	£15	£12	£8

No.217 Jaguar E Type.

Rich green with silver tints, headlights, bumpers etc. This model came from a great line of racing winners and was one of the most sought after in the whole Spot-On range. Complete with opening bonnet. Price 6/11d. Issued 1962. Deleted 1966. 106mm.

	MB	MU	GC
	£20	£15	£10
Also in red or scarlet, otherwise as above.	£15	£12	£10

No.219 Austin Healey Sprite.

An open sports car with driver. Silver grey or fawn with full silver trim etc. Price 4/6d. Issued 1962. Deleted 1966. 83mm.

	MB	MU	GC
	£15	£10	£8
Also in mustard or dark yellow with driver, passenger and red interior, otherwise as above.	£12	£9	£7

No.220 Austin Healey Saloon.

Yellow with full silver trim, headlights bumpers etc. Complete with driver and passenger. Price 5/6d. Issued 1963. Deleted 1966. mm.

	MB	MU	GC
	£20	£15	£10
Also in green, otherwise as above.	£15	£10	£8

MODEL	MB	MU	G

No.260 The Royal Rolls Royce.

One of the rarest and most sought after models in the car series. In royal maroon with silver trim, headlights, bumpers, grille and hubs. Has electric lighting in front and rear with four figures—Queen Elizabeth and Prince Philip, the Lady in Waiting and the Chauffeur. Price 19/11d. Issued 1964. Deleted 1966. 143mm. £50 £45 £3

No.261 Volvo P. 1800.

Light blue with silver tint, trim, headlights grille etc. and lemon or off-white interior with opening boot and bonnet. Price 6/6d. Issued 1963. Deleted 1966. 106mm. £15 £12 £9

No.261 Volvo P. 1800. Second Issue.

Yellow body with off-white interior and full silver trim. Price 6/11d. Issued 1965. Deleted 1966. 106mm. £10 £7 £5

Also with white body and rose pink interior otherwise as above. £7 £5 £3

No.262 Morris 1100.

Light or medium blue with pink interior. Opening bonnet, silver trim, headlights, bumpers and grille. Price 4/11d. Issued 1964. Deleted 1966. 88mm. £10 £8 £6

No.263 Supercharged Bentley.

Authentic Bentley green with silver headlights, spoked wheels and spare wheel on side, with folding windscreen and the number '27' and Union Jack decals. Price 13/6d. Issued 1963. Deleted 1966. 108mm. £30 £25 £2

Also in red, otherwise as above. £35 £30 £2

No.264 Armstrong Siddeley.

Silver or dark grey with full silver trim, headlights, grille and bumpers. Price 7/11d. Issued 1964. Deleted 1966. 106mm. £30 £25 £2

No.266 Bullnose Morris.

Made by one of the greatest companies of all times and a name that will live forever. Yellow with driver and opening dicky seat, red spoked wheels, black grille, running boards and bumpers with black hood and matching interior. Price 9/11d. Issued 1964. Deleted 1966. 85mm. £30 £25 £2

No.266 Bullnose Morris. Second Issue.

Sky blue and several in red which are are worth double. With black hood, interior, grille, running boards and mudguards, spare wheel on rear with driver and opening dicky seat. Price 9/11d. Issued 1964. Deleted 1966. 85mm. £50 £45 £3

No.267 MG 100.

Pink and grey with brown or gold interior, silver trim and opening bonnet. Price 5/11d. Issued 1965. Deleted 1967. 88mm. £15 £12 £9

Also in two-tone green, otherwise as above. Rare colour. £30 £25 £2

No.268 Model T Ford.

Blue with black hood and interior. Another rare model with spoked black wheels, gas tank and grille. Price 5/11d. Issued 1964. Deleted 1966. 88mm. £35 £30 £2

No.269 Ford Zephyr 6 and Caravan.

Ford is bright rose pink with white interior, passenger and female driver, with full silver trim, headlights, grille and bumpers. Caravan is in blue or medium blue with opening door, windows, red venetian blinds, white roof and opening side windows. Price 17/6d. Issued 1960. Deleted 1966. 275mm.

£35 £30 £25

No.270 Ford Zephyr 6.

Pale blue with red interior, female driver and poodle in back seat. With full silver trim, headlights, bumpers and grille. Price 7/11d. Issued 1964. Deleted 1967. 110mm.

£15 £12 £9

No.270 Ford Zephyr 6. Second Issue.

Maroon body with pale pink interior, female driver and poodle in back. New colour. Full silver trim. Price 7/11d. Issued 1975. Deleted 1977. 110mm.

£15 £12 £9

No.271 Riley Pathfinder.

Metallic silver with black and gold trim, silver headlights, grille, bumpers etc. Price 8/11d. Issued 1964. Deleted 1967. 110mm.

£30 £25 £20

No.274 Morris 1100 and Canoe.

Rose pink or medium red with matching pink and white canoe with paddles. Special roof-rack for boat and full silver trim, headlights, bumpers etc. Price 6/11d. Issued 1965. Deleted 1967. 98mm.

£15 £10 £7

No.276 S. Type Jaguar.

Silver metallic blue with red interior, silver trim, headlights, bumpers, grille etc. With driver and female passenger. Price 5/11d. Issued 1964. Deleted 1967. 113mm.

£15 £10 £7

Also in ruby metallic, otherwise as above.

£25 £20 £15

Also in cream with golden yellow top and interior. A very rare colour.

£50 £45 £35

No.278 Mercedes 230 SL.

Medium or light sky blue with red or pink interior, silver trim, hubs, grill, bumpers, headlights etc. Complete with driver. Price 5/11d. Issued 1965. Deleted 1967. 101mm.

£15 £10 £7

Also in ruby red with black roof, driver and passenger, otherwise as above.

£25 £20 £15

No.280 Vauxhall Cresta.

Purple with cream roof with white or silver side trims, pink interior, black steering wheel, silver trim, headlights, bumpers etc. Price 5/11d. Issued 1964. Deleted 1967. 110mm.

£15 £10 £7

Also in red with black roof and off-white or brown interior, otherwise as above.

£20 £15 £10

No.286 Austin 1800.

Blue or green with pink interior. One of the most reliable cars on the roads at a time when competition ran very high. Female driver and small boy passenger. Price 5/11d. Issued 1964. Deleted 1966. 100mm.

£15 £10 £7

Also in maroon, with silver white flashes.

£20 £15 £10

MODEL	MB	MU	G
No.287 Hillman Minx.			
Another beautiful and popular easy to run car. Rose red with white or blue interior, silver trim, headlights, bumpers etc. Price 5/6d. Issued 1964. Deleted 1966. 96mm.	£15	£10	£7
Also in two-tone blue, otherwise as above.	£20	£15	£1
Also in the rare colour of green with pink or lime interior, otherwise as above.	£30	£25	£2
No.288 Hillman Estate Car.			
Lemon with brown panels at sides and matching roof, full silver trim, headlights, bumpers etc. Price 6/11d. Issued 1963. Deleted 1967. 100mm.	£30	£25	£2
No.307 Volkswagen.			
Always advertised throughout the world as 'the last car on the scrap heap' because of its reliability. The first model was in rose pink with silver grille, headlights, bumpers and off-white or cream interior, complete with driver. Price 5/6d. Issued 1964. Deleted 1968. 110mm.	£15	£10	£7
Also in white or cream with cream or pink interior, otherwise as above.	£25	£20	£1
Also with blue body or the rare lime green. Seek the advice of an expert about rare colours.	£30	£25	£2
No.309 Ford Zephyr 6 Police Car.			
White with opening doors, police driver and companion, roof signs and 'Police' decals. Price 5/11d. Issued 1965. Deleted 1967. 110mm.	£20	£15	£10
Also in dark blue, otherwise as above.	£25	£20	£15

o.B23 Riva Speed Boat (Ariston).

assic, elegant, swift and constructed in accordance with the best
alian tradition. Easily manoeuvred with exceptional sea going
ualities. More solid than any other motor boat of the same period.
ch brown and golden yellow with orange seats, complete with
iver and full silver trim. Price 7/11d. Issued 1960. Deleted 1967.
5mm.

£20   £15   £10

o.B23 Riva Speed Boat. Second Issue.

he new design in brilliant red and white was made for the 1966
ternational Year and was sold at the Toy Exhibition in London
a very limited quantity. With matching red seats and white trim
es in a special presentation box which alone is worth £15. Price
6d. 155mm.

£50   £45   £30

o.CB106 Four wheeled Trailer.

pecially designed to be coupled onto either the Erf 68G or the
ammoth Major 8, which made a combination and an exact replica
the big road haulage unit. Bright yellow with white and off-
hite chassis. Matching white wheels. Price 4/11d. Issued 1960.
eleted 1967. 145mm.

£15   £10   £7

so in bright red or bright green with black chassis, otherwise as
ove.

£20   £15   £10

o.106A OC Austin Type 503.

etallic blue cab with matching trailer, a green container box with
ritish Motor Corporation' decals. The interior of the cab was
own or black with green tinted windows. Price 15/6d. Issued
60. Deleted 1966. 250mm.

£40   £35   £25

so in metallic red. Rare model.

£50   £45   £35

o.106A/1 Austin Type Lorry 503. Second Issue.

ustard or orange with matching trailer, side boards and tail-board,
ack chassis and tyres, metal wheels and silver trim, headlights and
umpers, brown or orange cab interior. Made in 1960 and deleted
1966. Price 17/6d. 150mm.

£35   £30   £25

so in metallic blue with matching trailer.

£40   £35   £20

so in dark red with matching trailer and dark cab interior,
herwise as above.

£50   £45   £35

o.108 Daimler Bus.

reen with silver trim and 'Continental Services Ltd.' on sides,
ening door, silver headlights, bumpers etc. Price 21/6d. Issued
63. Deleted 1966. 216mm.

£100   £75   £60

No.109 Erf Special Box Truck.

Red cab and chassis with large removal van body, black cab interior with opening doors at rear, complete with driver and mate. A rare model. Price 17/6d. Issued 1960. Deleted 1961. 220mm.

£75 £60 £

No.109/2 Erf Lorry 68G.

This modern lorry was the first to use the wraparound windscreen and was always popular with hundreds of operators. Because of its reliability, it represented one the Britain's major contributions to heavy goods transport. Lime green with matching chassis and float in metal grey. Black petrol tank, large tyres and silver grille, bumpers etc. Price 17/6d. Issued 1960. Deleted 1966. 220mm.

£50 £45 £

Also with red body and silver grey float, otherwise as above.

£75 £65 £

No.109/2/B Erf Lorry 68G. Second Issue.

Metallic light blue with silver grey float, metal wheels, black tyres and petrol tank, silver trim, headlights, bumpers, grille etc. Price 21/-. New design for 1962. Deleted 1967. 220mm.

£50 £45 £

No.109/3 Erf Flat Float with Sides.

This modern eight wheeler was king of its day, with its wraparound windscreen, mustard livery with brown interior and green tinted windows, full silver trim, headlights, bumpers, wheels etc. and black tyres. Price 17/6d. Issued 1960. Deleted 1966. 220mm.

£45 £40 £

No.110 Mammoth Major Cattle Truck.

Red cab and chassis with yellow cattle truck body, metal wheels, silver headlights and bumpers, black petrol tank. Complete with driver. Price 21/-. Issued 1960. Deleted 1966. 200mm.

£50 £45 £

No.110/2 AEC Mammoth Major.

This giant commercial vehicle was built by AEC of London, who made some of the best bus engines and chassis. It could haul cattle, bricks and even bulk liquids. Dark metallic blue with black petrol tank, orange cab interior with black steering wheel, silver grille, headlights, bumpers etc. Price 19/6d. Issued 1960. Deleted 1966. 200mm.

£50 £45 £

Also in rare colour of metallic red, otherwise as above.

£75 £65 £

No.110/2B Mammoth Major 8 Wagon.

Another fine vehicle by AEC in rose pink. Matching cab and float with 'London Brick Company Limited' on doors. Black petrol tank, silver grille, bumpers and trim. Price 17/6d. Issued 1960. Deleted 1966. 210mm.

£75 £65 £5

Also in metallic red with silver grille and the rare decals and emblem of 'British Road Services' in white lettering, otherwise as above.

£100 £85 £7

No.110/3 AEC Mammoth Major 8.

Metallic red with silver grille, metal wheels, and black tank. With 'British Road Services' decals on front of cab and vehicle sides in white lettering. Interior of open wagon is yellow. Price 15/11d. Issued 1961. Deleted 1966. 110mm.

£75 £65 £5

No.110/4 AEC Mammoth Major 8 Tanker.

Without doubt one of the most sought-after vehicles in the whole commercial range. Made by the makers of London's bus engines

MODEL	MB	MU	GC

and chassis. It had a 4,000 gallon fuel tank on a rigid chassis in the Shell livery, with red cab, off-white interior and lime green tanker body. 'Shell', 'BP' and 'Petroleum Products' emblems and decals in yellow, red, green and white on sides. Price 17/6d. Issued 1960. Deleted 1964. 210mm.

	£100	£85	£75

Also found with 'Esso' decals and badges in authentic red and white Esso livery.

	£100	£85	£75

Also in blue with matching tanker body with 'Blue-Col-Oils Limited'.

	£150	£125	£100

Also with yellow cab and dark green and white tanker body with 'American Oil Company' decals, white wall tyres. U.S. flag decals on front of cab. Also with 'Overseas Oil Company Route 66'. Both very rare.

	£250	£175	£150

No.111A/OT Ford Thames Trader.

Red with matching low loader body, green and dark mustard log load on rear. Price 14/11d. Issued 1960. Deleted 1966. 219mm.

	£75	£65	£50

Also with blue body, green trailer and pipe load, grey pipes. Otherwise as above.

	£75	£65	£50

No.111A/1. Ford Thames Trader.

Rose pink with lemon top cab and matching wagon interior. This model could be in other liveries and certainly had various names and companies connected with it, firstly 'British Railway' or 'Express Parcels Delivery'. Also models with 'G.W.R.' or 'L.N.E.R.'. Consult an expert about prices. Price 15/9d. Issued 1960. Deleted 1966. 219mm.

	£75	£65	£50

No.116 Caterpillar Tractor D9.

Mustard with silver blue blade and driver. One of the most impressive tracked vehicles ever produced. Often seen in action on many of the world's major construction companies. Powered by a turbo-charged 286 h.p. diesel engine, and equipped with a wide variety of attachments, the CAT D9 was capable sustained under the most arduous conditions. Bulldozer blade is raised and lowered by a powerful engine-driven winch. Price 15/6d. Issued 1960. Deleted 1966. 153mm.

	£50	£45	£35

Also in dark golden brown with blue blade and white tracks with white and blue driver, otherwise as above.

	£45	£40	£30

No.117 Jones Fleetmaster Crane.

One of the most versatile cranes ever built. This KL 10-10 was equipped with a single 125 h.p. diesel unit. It could travel on the road at speeds of up to 30 m.p.h. Completely mobile with a capacity load of 12½ tons. A great triumph for its makers K and L Steelfounders Limited, which belonged to the George Cohen 600 group. Red, grey and yellow with red and grey working hook. Yellow or mustard cab interior and blue tinted windows. 'Jones' decals in white. Price 19/11d. Issued 1960. Deleted 1966. 289mm.

	£40	£35	£25

No.118 Ford Thames Van.

These were nippy little servants and were used by such firms as Singer's Sewing Machines Limited. The van with this wording will be in lime green or dark green. Model in light blue has 'Hendees the Bakers' and 'Eat More Bread' and shows the picture of a loaf in the arms of a baker. Hendees was a famous company in the North east and their headquarters was based at Gateshead on Tyne. If anyone has any other liveries or unusual decals I would like to hear from them. Price 3/6d. Issued 1962. Deleted 1966. 79mm.

	£50	£45	£35

MODEL	MB	MU	GC

No.119 Morris Pick Up Truck.

Medium or dark green with silver headlights, bumpers, trim etc.
Price 3/6d. Issued 1963. Deleted 1966. 112mm.

	£25	£20	£15

No.112 United Dairies Milk Float.

Manufactured by United Dairies Limited in rich red ruby with
white roof, full milk load and 'United Dairy' decals. Cab has
sliding doors and driver. Price 5/6d. Issued 1960. Deleted 1966.
98mm.

	£45	£40	£25

No.122 United Dairies Milk Float.

Rose pink with cream roof, silver trim, 'United Dairies' decals in
gold lettering, driver and bottle load. Price 6/6d. Issued 1964.
Deleted 1967. 98mm.

	£35	£30	£20

No.122 United Dairies Milk Float.

Blue and white with 'Co-op' decals, driver and full silver trim.
Only a few of these were made in this livery. Price 7/6d. Issued
1965. Deleted almost at once. 98mm.

	£50	£45	£35

No.123 Bamforth Excavator.

The J.C.B. Hydra-Digga was made of tubular and boxed steel
section. Of robust construction, capable of withstanding all shocks
and stresses. Powered by a 51.8 h.p. engine and operated by five
Rams. Especially designed and produced by J.C.B. and fittd with
750 x 16 front and 14 x 30 rear wheels and tyres. Fitted with
electric lighting and starting, plus handbrake and all-weather
driver's cab. Bright yellow with Red Line design and 'J.C.B.' decals
and lettering in black on white back ground. Large red wheels and
thick tyres at rear, with smaller wheels at front. Red scoop and
shovel with white or yellow cab interior. Price 12/6d. Issued 1960.
Deleted 1967. 272mm.

	£30	£25	£20

No.123/A Bamforth Excavator, Second Issue

Orange and black with silver scoop and shovel. Large and small
black wheels. With 'Road Work' and 'Mighty-Mover' decals and
emblems. Rare. Price 15/-. New design for 1962. Deleted 1967.
272mm.

	£50	£45	£35

No.135 L.G.P. Sailing Dinghy.

Undoubtedly one of the best designs and most successful ever
made. The G.P.14 was made suitable for racing and rowing and for
use with an outboard motor. Designed by Jack Holt. Could be
bought as a complete boat or as a kit. Blue and white with silver or
metal trailer with metal wheels and black tyres. Price 6/11d. Issued
1960. Deleted 1966. 117mm.

	£15	£10	£7

Also in red and white with black trailer, otherwise as above.

	£25	£20	£15

No.137 Massey Harris Tractor.

This popular tractor had no equal and it boasted an efficient
hydraulic system for raising and lowering various implements.
Bright red with silver grille, blue metallic wheels and matching
engine, 'Massey Ferguson' decals, driver and white steering wheel.
Driver has a yellow jumper and purple trousers, white collar and
red tie. Designed to pull a range of Spot-On implements. Price
8/6d. Issued 1960. Deleted 1966. 79mm.

	£25	£20	£15

Also in green with black wheels and tyres, 'Massey Harris' decals
and driver in brown and white, otherwise as above.

	£20	£15	£10

MODEL	MB	MU	GC

No.138/1 Combine Harvester.

Red, yellow and black with two drivers. Price 22/11d. Issued 1960.
Deleted 1962. 246mm.

£50 £45 £35

No.139 Eccles E16 Caravan.

Apart from the Eccles name being world famous, the E16 was one
of the most popular caravans ever built. A regular sight on almost
every coastline. Very comfortably furnished, it provided many
happy holidays for its owners and their families. Blue with white
roof, metal wheels, silver hubs and metal tow bar. With orange and
white dotted curtains and opening door. Price 9/11d. Issued 1961.
Deleted 1966. 146mm.

£25 £20 £15

Also in white with blue outline, opening door and windows,
otherwise as above.

£30 £25 £20

Also in the rare colour of red, with cream roof and matching
wheels, otherwise as above.

£50 £45 £35

No.145 London Routemaster Bus.

This magnificent vehicle is the original service car that took the eye
of tourists and the world by storm. In the Greater London area,
thousands were in every day use and the only original Routemaster
to be seen is now in the hands of bus enthusiasts and museum
keepers. Lineage goes back to the pre-1914 horse drawn omnibus.
The first model was in red with 'Pedigree the Finest Baby
Carriage' decals on sides and 'Ninety Three Putney Triang Toys
for Girls and Boys' on front and rear. Driver with blue uniform.
Black grille, silver headlights and blue tinted windows. Price
9/11d. Issued 1959. Deleted 1966. 198mm.

£150 £125 £100

Also in the promotional green and cream livery. Very rare. Made
specially for the 1966. International Toy Year. In special
presentation box. Price 21/-. 198mm.

£250 £200 £150

No.156 Mulliner Luxury Coach.

Modern luxury craftsmanship at its very best. It boasted advanced
body styling plus the last word in comfort for its many passengers
who were lucky enough to travel inside it. It was a far cry from the
old English stage coach. This bus, like many other models of this

type, are scarce. They will be almost unobtainable in the future and one of the best investments. Perhaps you may think that the price I have quoted is high, but this is because of the great demand for the models. Soon the only place that you will be able to see a Spot-On model of this type will be in a museum. Every kind of bus is now being sought after, so if you have any of these or any of the heavy goods vehicles, make sure that you get the correct price for them. You can always write to me for advice, or consult an expert at some collector's shop. Silver metallic grey on the upper part of bus body and dark rich metallic blue on the lower. A brown chev design with a thin red line runs underneath the chev. Silver grey grille, headlights, bumpers and window trim, LMC 156 number plates, dark brown seats with silver trimming. Fibre glass roof, silver metal wheels with fancy design and large rubber tyres. Used extensively by private collectors and show organizers as a promotional model. Several adverts and signs are connected with the model and other liveries do exist. Price 17/11d. Issued 1960. Deleted 1967. 207mm approx.

£150   £125   £1(

No.158/A 3C Bedford Low Loader.

An all-purpose general vehicle. Could be obtained as an articulated or rigid chassis version. An articulated prime mover fitted with Scammell coupling gear, hauling a low loader with cable drum winch unit in rose red, blue wheels with large grey tyres and black petrol tank. Price 21/-. Issued 1961. Deleted 1966. 207mm.

£50   £45   £3(

No.158/A/02 Bedford 10 Tonner.

This powerful diesel engined vehicle can be obtained as an articulated or rigid chassis version. This model is an articulated prime mover fitted with Scammell coupling gear, hauling a Shell 2,000 gallon tanker. Green cab with 'Shell' decals and distinctive designs. Red tanker body with 'Shell BP' decals, black mudguards, silver grille, headlights, bumpers, black petrol tank and grey cab interior with clear windows. Price 22/6d. Issued 1960. Deleted 1967. 199mm.

£50   £45   £3(

No.161 Long Wheel Base Land Rover.

A special variation of the well-known Land Rover of which more that 250,000 have been produced. It had a rugged construction, a powerful engine and four-wheel drive which enabled it to go almost anywhere. This tropical version has a double roof for heat protection. Grey or silver grey with cream or lemon roof. Large metal wheels and thick special tyres. Price 12/6d. Issued 1961. Deleted 1967. 108mm.

£40   £35   £2(

Also in medium or dark blue, otherwise as above.

£50   £45   £3(

No.207 Wadham Ambulance.

This vehicle had an all fibre glass body with the best fitting connected to a BMC petrol or diesel engine and chassis unit, with wraparound windscreen and an ingenious interior arrangement. A fine example of British specialized vehicle design. Lemon and lime with silver blue trim, grille, headlights, bumpers, wheels etc. Large black tyres and white interior, blue tinted windows, opening rear doors, 'Ambulance' sign in front. Price 10/6d. Issued 1961. Deleted 1967. 108mm.

£40   £35   £2(

MODEL	MB	MU	GC

No.207 Wadham Ambulance. Second Issue.

White body with full silver trim, headlights, bumpers, grille, wheels
etc. With black steering wheel, driver, 'L.C.C.' decals and
ambulance sign on front. Price 12/6d. New design for 1962.
Deleted 1967. 108mm.

	MB	MU	GC
	£30	£25	£20

Also in dark blue with light blue trim, otherwise as above. £40 £35 £25

No.210 Morris Mini Minor Van.

This model was so much in demand that the new livery and design
of dark blue was issued in 1964 and deleted in 1967. Price 4/11d.
70mm.

	MB	MU	GC
	£15	£12	£9

No.210 Morris Mini Minor Van. Third Issue.

Dark blue with matching set of wheels, 'British Tyre Company'
decals and Union Jack design on doors and sides. Price 5/6d. Issued
1965. Deleted 1968. 80mm.

	MB	MU	GC
	£50	£45	£35

Also in red with matching wheels and 'Crawford's Biscuits Are
Good Biscuits' on sides, otherwise as above. £50 £45 £35

Also in gold or dark yellow with matching wheels, a jam jar
emblem on doors and sides and 'Golden Shred' on sides. £75 £65 £50

Also in light blue or medium blue with white lines and black
wheels with grey tyres. With 'Dewhurst's Argentine Beef' decals in
gold lettering, with a bull's head design on doors and sides. A rare
promotional model. £100 £85 £75

No.264 Tourist Caravan.

One of the more specialized products and is another rare item to
find. Light blue, with red centre line all round model and white
roof. Fully fitted inside with red curtains and opening door. Black
tow-bar, silver hubs and wheels with black tyres. White interior.
Price 9/6d. Issued 1965. Deleted 1967. 164mm.

	MB	MU	GC
	£30	£25	£20

Also in red, otherwise as above. £50 £45 £35

Also in white with blue line, otherwise as above. £15 £12 £7

Also in lime green with matching lines and wheels, grey tyres and
matching curtains. £50 £45 £35

No.265 Tonibell Ice Cream Van.

Another Spot-On classic model worth its weight in gold. Light blue
with orange red thick line design along sides. 'Tonibell' decals on
sides and front, 'Soft Ice Cream' decal on front and rear. Ice cream
server in white and white van interior. Silver headlights, grille, trim
etc. Price 8/11d. Issued 1964. Deleted 1967. 107mm. £50 £45 £35

Also in metallic red and cream with golden lines, matching grille,
bumpers and headlights. Ice cream server in light blue coat,
otherwise as above. £100 £85 £75

Also in cream with red line decal. £75 £65 £50

Also in two-tone blue with 'Mr. Softee' decals. Very rare. £150 £125 £100

No.269 Zephyr 6 and Classic Caravan. Second Issue.

Deep red with black trim lines, both on Zephyr and caravan. Price
21/-. A new design made for 1965. Deleted 1968. 275mm. £50 £45 £35

Also in white with black lines, matching caravan, grey tyres. £100 £85 £75

No.271 Express Dairies Milk Float.

Another prize model with a great following. This model was soon
snapped up after being placed on sale. Dark blue with white milk
bottle load. Blue driver, with 'Express Dairies' decal and design on
front. 'Drink Express Milk' decals on sides at top. White roof on
float, black wheels. Price 6/6d. Issued 1964. Deleted 1968. 98mm. £30 £25 £2

Also in red with blue roof and driver in white with 'Drink More
Milk' decals, otherwise as above. £35 £30 £2

Also in two-tone blue with white roof and 'Co-op' decals. £50 £45 £3

No.273 Commer Security Van.

Although the double decker bus and the Mulliner coach are very
rare, an even rarer model was the money box Commer Security
Van. Dark or medium green with 'Security Express Limited' on
sides in yellow or gold lettering, complete with emblem. Driver in
blue uniform and hat. With opening rear doors, interior cage with
security guard lock, silver grille, lights, bumpers and wheels with
black tyres. Price 12/11d. Issued 1965. Deleted 1967. 125mm. £150 £100 £5

Also in black or dark blue with 'Securicor' decals, otherwise as
above. £250 £150 £1

No.308 Land Rover and Trailer.

Olive green with fawn canvas back on Land Rover, metal grille,
headlights and hubs, thick black tyres, silver bumpers, black
steering wheel and dark green cab interior. Price 17/6d. Issued
1965. Deleted 1967. 255mm. £30 £25 £2

Also in medium or dark blue, with grey tyres, otherwise as above. £50 £45 £3

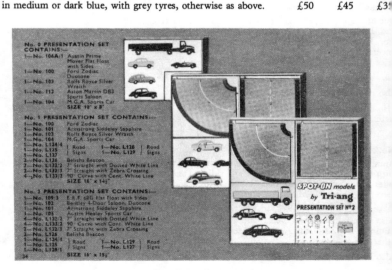

PRESENTATION SETS

1.O Set.

The first of a limited number of superb presentation sets. The company did not last for long and sets are extremely rare, especially the early ones. Contains Austin Prime mover; flat float with sides; Ford Zodiac; Rolls Royce Silver Wraith. Aston Martin D.C.3 and MGA Sports car. Colours may vary on models in all sets. Price 37/-. Issued 1960. Deleted 1966. Box 255 x 204mm.   £100

1.0A Set.

Contains Austin Taxi Cab, Goggomobil, Super Bristol 406, Jaguar XSS, Rover 3 litre, Fiat Multipla, Erf 68G with flat float. Complete with white lines, crossings and belisha beacons. Price 37/-. Issued 1961. Deleted 1967. Box 286 x 325mm.   £75

1.1 Set.

Ford Zodiac, Armstrong Siddeley Sapphire, Rolls Royce Silver Wraith, MGA Sports car, road signs, zebra crossing and white lines. Price 42/6d. Issued 1960. Deleted 1966. Box 408 x 370mm.   £75

1.2 Set.

Contains ERF 68G Flat float with sides, Bentley 4-door Saloon, Austin Healey Sports car, road signs, crossings etc. Price 45/-. Issued 1960. Deleted 1966. Box 408 x 395mm.   £75

1.2A Set.

Contains Bedford 10 tonner, 2,000 gallon tanker, LWB Land Rover, Austin Taxi, Austin Healey, Meadows Frisky Sports Car, Vauxhall Cresta, Rover 3 litre, ERF 68G flat truck, flat float with sides. Price 57/6d. Issued 1961. Deleted 1966. Box 312 x 395mm.   £100

1.3 Set.

Contains Thames Trader with Artic., flat float, Armstrong Siddeley Sapphire, MGA Sports Car, Triumph TR3, Jensen 541, Aston Martin DB3, Jaguar 3.4., beacon, straights and road signs. Price 37/-. Issued 1961. Deleted 1966. Box 408 x 433mm.   £100

1.3 Set. Second Issue.

Contains Mulliner coach, Bedford 10-ton tanker, Eccles caravan, Wadham ambulance, road signs and beacons. Price 66/6d. Issued 1961. Deleted 1962.   £200

1.4 Set.

Contains Austin Prime mover with flat float, Erf 68G with flat float, Rolls Royce Silver Wraith, Ford Zodiac, Jensen 542, Jaguar XS, straight white lines, road signals and belisha beacons. Price 37/-. Issued 1961. Deleted 1966. Box 408 x 433mm.   £100

1.4A Set.

Contains BP filling station, straights, dotted lines, zebra crossings, road signs, plus 'Road Up' or 'Under Construction' signs, LWB Land Rover, Sunbeam Hardtop, Vauxhall Cresta, AEG Major 8 with brick load. Price 56/-. Issued 1960. Deleted 1966. Box 542 x 383mm. approx.   £100

1.702 Set.

Contains a Ford Zephyr 6 with opening doors; Morris 1100 with roof-rack, canoe and opening bonnet; Austin 1800 with opening bonnet and sailing dinghy on trailer. Price 25/11d. Issued 1960. Deleted 1966. Box 408 x 403mm.   £50

No.260 The Royal Presentation Set.

One of the most sought-after sets in the whole of the Spot-On range. Often talked about but seldom seen. Contains a royal Rolls Royce in royal purple with full silver trim, headlights, bumpers, grille and hubs. Royal figures of Queen Elizabeth and Prince Philip, a lady in waiting and the driver in his splendid uniform. Two guardsmen and several figures to use in royal procedures. Car has opening boot, electric front and rear lights and interior light. Price 30/6d. Issued 1964. Deleted 1967.   £100

No.0 set. Second Issue.

Contains colour variations among the models. Some are rare. Numbers: 100, 103, 104, 106A/1 and 113. Price 75/-. Issued 1963. Deleted 1967.   £100

No.1 Set. Second Issue.

Another set with colour variations. Contains numbers: 100, 101, 103, 104, L124/4, L125, L127, two of number L26, two of L132/2, two of L132/3 and four of L133/2. Price 64/-. Issued 1964. Deleted 1967.   £100

No.2 Set. Second Issue.

Contains an assortment of curves, lines and other road accessories, plus one model each of numbers: 101, 102, 105, 109/3, L125 and L128/1. Price 65/-. Issued 1965. Deleted 1967.   £75

No.3 Set. Second Issue.

Contains the following road signs, Numbers L124/1, L124/2, L124/3, L124/4, L124/5, L124/13, L125, L127/1, L127/2, L128/1, L128/2 and L128/3. Also contains one of each of the following models: numbers 101, 104, 108, 111A/1, 112, 113 and 114. Also four of each of the following numbers: L32/3, L33/2, L132/2 and L132/3. Price 75/-. Issued 1962. Deleted 1967.   £125

No.4 Set. Second Issue.

Contains an assortment of lines, curves, beacons etc, plus one of each of the following models: numbers 100, 103, 106/1, 107, 109/3, 112, L124/1, L124/2, L124/4, L124/5, L125, L127/1, L127/2, L128/1, L128/3 and L172. Price 85/-. Issued 1963. Deleted 1966.   £200

No.0A Set, Second Issue.

Contains one of each of the following models: 107, 109/2, 115, 120, 131, 155 and 157. Price 57/6d. Issued 1964. Deleted 1966.   £75

No.2A Set. Second Issue.

This set may have models which vary in colour and they could be worth a good deal more than the normal livery. Always consult an expert. Set contains one of each of the following models. 105, 109/3, 119, 155, 157, 158A/2, 161 and 165. Price 57/6d. Issued 1965. Deleted 1967.   £75

No.2AA Set.

Contains a number 2 garage and one of each of the following models: numbers 145, 156 in blue and silver and red and cream. Price 75/-. Issued 1965. Deleted 1967.   £250

No.4A Set. Second Issue.

Contains one of each of the following numbers 110/2B, 131, 154, 161, 162/B, 165 and 191/1, plus five road signs: two number L132/3, four L132/2s and four L133/2s plus two belisha beacons. Price 55/-. Issued 1964. Deleted 1967.   £50

No.L208. Road Construction Set. (A).
Contains four workmen, one workman's hut, five roadworks signs, one stop and 'go' sign, one workman's hole. With imitation soil, four warning lamps and barriers. Price 7/6d. Issued 1960. Deleted 1966.   £10

No.L208 Road Construction Set. (B).
Contains four night warning lamps, four barrier poles, four barrier pole standards, a watchman's hut, a go standard, a workman's fire, danger signal, 'Road Works Ahead' sign, 'Road Up' sign and two figures in a well-packed beautifully illustrated box. Price 7/6d. Issued 1964. Deleted 1967.   £10

No.L162 Filling Station Set.
A perfect scale model of a typical filling station and could be obtained as L162/A with 'Shell' decals or as L162/B with 'BP' decals. Each part fitted easily together and was very realistic. Price 2/6. Issued 1964. Deleted 1967.   £10

No.L162/1 Garage Set.
Another fine service station strongly made with a metal base, finished in either the 'Shell BP Company' or the rare 'Esso Company' which is worth double. Price 12/6d. Issued 1964. Deleted 1966.   £10

No.2L162/2 Garage.
Another strong service station with a steel base. With Shell, 'B.P.' or the rare 'Total' decals, worth double, in a very attractive presentation box. Price 15/-. Issued 1965. Deleted 1967.   £15

No.3L162/3 Garage Set.
This station and garage was one of the larger types and will be very hard to find in mint condition. Complete with opening doors, pumps, oil dispensers, management and staff. Price 21/-. Issued 1965. Deleted 1967.   £50

No.0 Master Roadway Set.
Contains one of each of the following: 101, 102, 103, 104, 105, L33/2, L132/1, L132/2, L132/3, L133/1, L152/1, L152/2, L153, 175, L176, L177, L178, L182 and L204, contains set of personnel figures, station attendants, drivers, office girl and manager with Land Rover. Price 95/-. Issued 1964. Deleted 1966.   £250

No.1 Master Roadway Set.
Contains one each of the following numbers: 107, 108, 112, 113, 114, 115, L132/1, L132/2, L132/3, L133/1, L133/2, L152/1, L152/2, L153, 173, L175, L176, L177, L178, L182 and L204 and a set of figures and seats. Price 95/-. Issued 1965. Deleted 1967.   £75

No.702 Second Issue.
Contains a Ford Zephyr 6 in blue with opening doors and driver, a Morris 1100 in green with roof-rack and green and white canoe, an Austin 1800 in red and black with matching dinghy on trailer and a tourist caravan in red, blue and black. Very rare set. Price 21/-. Issued 1965. Deleted 1967.   £250

Roadways Accessory Set.
A rare set in the boxed gift series and contains one of each of the following: L121, L124/1, L124/2, L124/3, L124/4, L124/5, L124/6, L124/7, L124/8, L124/9, L124/10, L124/11, L125, L126, L127/1, L127/2, L127/3, L127/4, L127/5, L127/6, L128/1, L128/2, L128/3,

L129, L130/1, L130/2, L130/3, L134, L144, L151, L160, L172/A,
L172/B, L205 and a box of figures and various trees etc. One
special Jones 117 crane; one number 123 excavator and a Terrapin
building set. Price 72/6d. Issued 1965. Deleted 1967.   £100

Special Milk Vans Set.

This rare set contains one of each of the following. Express Dairies
Milk Float with driver, United Dairies Milk Float with driver, dual
purpose ice van for ice cream or normal milk runs, a 10-ton milk
tanker in white and blue and one Erf flat float with sides and milk
churn load. Price 65/-. Issued 1965. Deleted 1966.   £250

Special Farm Gift Set.

Another rare set contains one of each of the following: 116, 137,
161, 110/2B, 111A/OT, CB106 and Thames Trader. Complete
with farm animals, bales and fencing. Price 50/-. Issued 1965.
Deleted 1966.   £100

National & International Registration Plates.
Morgan, Humber, Sunbeam, Jaguar, Bentley, Aston Martin, Morris, Allard, Frazer Nash, Daimler
Lion, Hillman, Armstrong Siddeley, Rolls Royce, Standard, Triumph, Rover, Riley, Alvis, M.G
Lotus, Vauxhall, S.M., Turner, Wolseley, Bristol, ERF, Elva, Berkeley, AC, Ford, Mercedes
Peugeot, Morgan-Two, Fiat, Chrysler, Chevrolet, Citroen, Buick.

A. Austria, L. Luxembourg, SF. Finland, D. Germany, DK. Denmark, PE. Peru, H. Hungary, R
Lebanon, B. Belgium, N. Norway, CH. Switzerland, I. Italy, IR, Iran, AUS. Australia, S. Swede
F. France, EIR. Eire, TR. Turkey, E. Spain, IRQ. Iraq, IL. Israel, NZ. New Zealand, MC. Monac
GR. Greece, USA. United States of America.

The Set of British Road Signs
Two children holding hands. Sign of a plane, 'Low Flying aircraft or Noise'. Car and Curling Line
'Slippery Road'. No Right Turn, No Left Turn, No U.Turn, Steep Hill Upwards, Falling Roc
Double Bend, Max. Speed Limited 30. End of Max, Speed Limited. Stop Police. Cattle. Two w
Traffic Ahead. Uneven Road. No Waiting. No Stopping Clearway. No Entry. One way Traffi
Advance warning of No Through Road. No Overtaking. No Cycling or Moped Riding. Cyclists ar
Moped Riders only. Pass Either Side. Belisha Beacons. Lamposts, Telegraph Poles, etc.

Any other information will be entered in the Price Guide section and trust that the Spot-On story h
given you an idea of what went into making such a marvellous range of products.

Prices for the Registration plates and badges will vary from £1, up to £10 each according
availability. These must be in mint condition to gain the top prices, although I would imagine ar
keen collector of these items will pay money for them regardless of condition.

TRI-ANG-MINIC SHIPS

INTRODUCTION

This book would not be complete without the fine die-cast Minic ships which were introduced in 1951 and were still being made in 1954, although the deletion dates are questionable. Ships were still found in shops in the early 1970s and it is quite possible that there are still a few hidden away in store-rooms at the present time. Some of them will be certainly stored away as good investments, while others will be just lying on a shelf or packed in a box unknown to the person who actually owns them.

The ships belong to a special collector's category, as the people who collect ships certainly do not collect model cars. Owing to the demand of these beautifully made models, the Triang Company made several thousand, although their actual named models were limited. The prize ships in this range without a doubt are the liners like the *Queen Elizabeth*, the *Queen Mary*, the *United States* and the very rare *Aquitania*.

I have listed every model as made in 1951/54 and deleted in 1970. These dates are authentic, and as the range of ships is limited it will meet the needs of the ship-collecting specialist. Should anyone have any queries or need advice and personal valuations of items belonging to this section, I shall be pleased to hear from them through the kindness of may publisher. Please enclose a stamped addressed envelope which will ensure you a definite reply.

BOXES

All of these models were packed in the yellow, cream or white boxes with black lettering. Apart from the ships, there were accessories such as buildings, lighthouses and other things connected with the sea. I have not included these in the list, as I found that they were not actually collectable - the ships themselves are the real investments.

MODEL	MB	MU	GC
M702 Queen Elizabeth One of the finest liners ever to leave the Clyde and certainly the pride of the Scottish shipbuilding industry. The model itself is quite rare and one of the most sought-after in the entire range. In black and white or black and cream with red and black or red, black and blue funnels. Price 3/6d. 250mm. approx.	£25	£15	£10
M703 Queen Mary A magnificent liner which has been admired in every part of the world. Black and white or black and cream. Black and red funnels with black liners. Price 3/6d. 255mm.	£25	£15	£10
M704 S.S. United States of America Another magnificent liner reputed to be one of the largest passenger ships ever to sail the high seas. A very fast ship decked out in black and white, complete with 24 lifeboats ready to drop at a minute's notice. The funnels are in red, white and blue. Price 3/6d. 250mm.	£25	£15	£10
M705 R.M.S. Aquitania A unique model in white, brown and black with four red funnels. Complete with 26 lifeboats. Price 3/6d. 230mm.	£20	£12	£9
M708 R.M.S. Saxonia White and black with one red and black funnel, complete with 12 lifeboats. Price 2/6d. 150mm.	£15	£10	£5
M714 Flandre Black and white with one red and black funnel, complete with 12 large lifeboats and two smaller lifeboats. Price 2/-. 150mm.	£15	£10	£5
M716 Port Brisbane Grey and white with one black and red striped funnel. Complete with four lifeboats. Price 1/11d. 140mm.	£9	£7	£4

MODEL	MB	MU	G
M721 The Royal Yacht Britannia			
This famous liner was used as a Red Cross hospital ship during the war. All-white livery with one funnel showing the red cross. Price 1/6d. 100mm.	£15	£10	£.
M723 The Isle of Guernsey			
Black and white with two yellow and black funnels, complete with eight lifeboats. Price 1/-. 75mm.	£5	£3	£
M731 Tug			
Blue, white and brown with a black base and red and black funnel. These little tugs were life-savers to the huge liners which could not have survived without their assistance, especially in bad weather and thick fog. Price 6d. 40mm.	£1	75p	5
M732 S.S Varicella			
A very useful craft which could carry all types of heavy goods including coal, oil and various types of steel. White and black with a yellow and black funnel. Several of these were used by the Shell Oil Company and some models show their colours and emblems. Price 1/6d. 165mm.	£7	£5	£.
M741 H.M.S. Vanguard			
One of the largest and finest battle ships ever to serve in the Royal Navy. Complete with revolving guns which were unique on such a small die-cast model. In grey wartime livery. Price 2/6d. 200mm.	£15	£10	£5
M761 H.M.S. Swiftsure			
Another fine model of a warship with a great name and history. Price 1/11d. 145mm.	£7	£5	£3
M771 H.M.S. Daring			
Another model of a fighting ship in wartime grey and also in light blue, which is worth double. Price 1/-. 100mm.	£7	£5	£3
M772 H.M.S. Diana			
Grey or light blue, worth double. Price 1/-. 100mm.	£7	£5	£3
M773 H.M.S. Dainty			
Grey or light blue, worth double. Price 1/-. 100mm.	£7	£5	£3
M782 H.M.S. Tobruk			
A famous name and an equally famous model in battle ship grey. Price 1/-. 100mm.	£7	£5	£3
M783 H.M.S. Hampshire			
A fine model of a fighting warship complete with helicopter and rotating gun. Price 2/-. 140mm.	£8	£6	£4
M789 H.M.S. Virago			
In battle ship grey, this is another model of a fine fighting warship. Price 1/-. 90mm.	£4	£3	£1
M790 H.M.S. Volage			
Blue or battle ship grey, worth double. Price 1/-. 90mm.	£4	£3	£1
M793 H.M.S. Blackpool			
Blue or battle ship grey, worth double. Price 1/-. 90mm.	£4	£3	£1

MODEL	MB	MU	GC
M794 H.M.S. Tenby			
Blue or battle ship grey, worth double. Price 1/-. 90mm.	£4	£3	£1
M799 H.M.S. Repton			
Battle ship blue with blue and black funnel. Sold in boxes of six and twelve. Price for individual model 6d. 40mm.	£1	75p	50p
Price for box of six 3/-	£5	£4	£3
Price for box of twelve 5/-	£10	£8	£6
M810 H.M.S. Turmoil			
Wartime grey with black and grey funnel. Sold in boxes of six or twelve. Price 3/-. 50mm. approx	£5	£4	£3
Price for box of twelve 5/-	£10	£8	£6
Submarine			
Wartime grey, sold in boxes of six or twelve. Price 3/-. 70mm.	£5	£4	£3
Price for box of twelve 5/-	£10	£8	£6

EMPTY BOXES, CATALOGUES AND LEAFLETS

As in previous books which I have written, I cannot stress the value of empty boxes. A box makes all the difference to a model, especially when trying to sell them.

Every authentic collector likes to buy a 'Boxed' toy or model whenever possible. Boxes have a psychological effect, and a Minic model in a box is always worth considerable more than an unboxed one. I have dealt in the selling of empty boxes since 1976. As an example, a No.29M Minic patrol car without a box in worth £12 while with a box it is worth £15. This makes the empty box worth £3.

A box can help to date and authenticate a model, and a box with a picture on it could be worth a lot more than a plain one. The prices I quote for empty boxes are quite realistic and I see that in the very near future many shops will open in all the major towns and cities with a section devoted solely to boxes and catalogues.

Triang Railways	Valu
R50 Princess Elizabeth	£
R52 Tank Loco	£
R53 Princess Elizabeth	£
R54 Pacific Loco	£
R55 Diesel Loco	£
R56 Electric Tank Loco	£
R55 Canadian National Diesel	£
R57 Diesel Dummy End	£
R58 Diesel B Unit	£
R59 Tank Loco	£
R138 Snow Plough	£
R150 Class B12 Locomotive	£
R151 Saddle Tank Loco	£
R152 Diesel Shunter	£
R153 Saddle Tank Loco	£
R154 Diesel Shunter	£
R155 Diesel Switcher	£
R156 Suburban Motor Coach	£
R157 Diesel Power Car	£
R158 Diesel Trailer Car	£
R159 Double Ended Diesel	£1
R159/B Double Ended Diesel	£
R225 Suburban Motor Coach	£
R251 3F Loco	£
R253 Dock Shunter	£
R254 Steeple Cab Loco	£
R255 Saddle Tank Loco	£
R257 Double Ended Electric Loco	£
R258/S Princess Royal	£
R259/S Britannia	£
R346 Stephenson's Rocket Train	£
R350 Class 3P Loco	£
R351 Co-Co Class Electric Loco	£
R352 Budd Diesel Car	£
R353 Yard Switcher	£
R354 Lord of the Isles	£
R355 Connie Loco	£
R355ER Polly Loco	£
R355/Y Connie Loco	£
R356 The Battle of Britain Loco	£
R357 Diesel Loco	£
R358 Davy Crockett Loco	£
R359 Tank Loco	£
R386 Princess Elizabeth	£
R550 Saddle Tank Loco, Clockwork	£
R553 Caledonian	£
R555 Diesel Pullman Car	£
R556 Non-powered Pullman Car	£

559 Diesel Loco	£3
644 Bo-Bo Electric Loco	£4
645 Hynek Diesel Loco	£3
653 Prairie Tank Loco	£5
751 Electric Co-Co- Diesel	£5
752 Battle Space Turbo Car	£5
753 E3 Class Electric Loco	£5
754 M7 Class Tank Loco	£5
758 Hymek Hydraulic Loco	£3
759 Albert Hall Loco	£5
850 The Flying Scotsman	£5
W 2218 4 MT Class Loco	£5
W 2224 AF Loco with Tender	£5
W 2226 City of London Loco	£5
W 2233 Co-Bo Diesel Electric Loco	£5
W 2235 West Country Barnstaple Loco	£5
W 2245 .3002 Electric Loco	£5
W 2250 Electric Motor Coach	£5
W 4150 Electric Driving Trailer Coach	£3
W 2207 Tank Loco	£3
W 2217 Tank Loco	£5
W 2231 Diesel Electric Shunter	£2
W 397 The Satellite Train	£5
W 398 Strike Force 10 Train	£5
20 LMS Coach	£2
21 BR Coach	£2
22 SR Coach	£2
23 Royal Mail Coach	£3
24 Silver Coach	£2
25 Vista Dome Coach	£3
26 LNER Coach	£3
27 Pullman Coach	£5
28 Main Line Brake Third Coach	£2
R29 BR Main Line Composite Coach	£2
111 Hopper Car	£2
114 Box Car	£1
115 Caboose	£2
119 Mail Coach	£5
119 Mail Coach	£2
120 BR Coach	£2
121 Suburban Composite Coach	£2
123 Horse-Box	£2
125 Observation Coach	£1
130 Baggage Car	£1
131 Coach	£2
132 Vista Dome Car	£2
133 Observation Car	£2
134 Baggage Car	£2
220 Main Line Brake 3rd Coach	£3
221 Main Line Composite Coach	£3
222 Suburban Brake 3rd Coach	£3
223 Suburban Composite Coach	£3
224 Restaurant Car	£3
225 Main Line Coach	£2
226 Utility Van	£2
226B Utility Van	£2
227 Utility Van	£2
228 Pullman 1st Class Car	£3
230 Coach	£2
231 Coach	£2
248 Ambulance Car	£3
230 BR Brake 2nd Coach	£2
321 BR Composite Coach	£2
322 Restaurant Car	£2

R324 Diner
R325 Diner
R328 Pullman Brake 2nd Car
R329 Brake 2nd Coach
R330 WR Composite Coach
R331 WR Restaurant Car
R332 GWR Composite Coach
R333 GWR Brake 3rd Coach
R334 Diesel Rail Car Unit
R335 Coach
R336 Observation Car
R337 Baggage/Kitchen Car
R338 Dining Car
R339 Second Class Sleeping Car
R382 Composite Coach
R383 Brake 2nd Coach
R400 Transcontinental Mail Coach
R401 Transcontinental Mail Coach
R402 Royal Mail Coach
R422 BR 2nd Composite Coach
R423 BR Brake 2nd Coach
R424 Buffet Car
R425 Parcels Brake Coach
R426 Pullman Parlour Car
R427 Caledonian 1st/3rd Composite Coach. Rare
R428 Caledonian Brake Composite Coach
R429 Caledonian 1st/3rd Composite Coach
R440 Transcontinental Coach
R441 Observation Car
R442 Baggage Kitchen Car
R443 Transcontinental Diner
R444 Continental Coach
R444/CN Canadian National Passenger Car
R445 Observation Car
R445/CN Canadian National Observation Car
R446 Baggage Kitchen Car
R447 Diner Transcontinental
R448 Old Tyme Coach
R620 Engineering Coach
R621 Liverpool/Manchester Coach
R622 Main Line Coach
R623 Main Line Brake 2nd Coach
R624 Buffet Car
R625 Continental Sleeper Car
R626 Main Line Coach
R627 Main Line Brake 2nd Coach
R628 Main Line Buffet Car
R722 Inter City 2nd Class Coach
R723 Inter City Brake 1st Coach
R724 Inter City 2nd Class Coach
R725 Dining Car
R726 Inter City Brake 2nd Coach
R727 Blue Composite Coach
R728 Blue Brake Coach 2nd
R729 Blue Buffet Car
R730 Composite Coach
R731 Blue Brake 2nd Coach
R732 Blue Buffet Car

VANS AND WAGONS

R10 Goods truck	£1
R11 Goods Van	£1
R12 Shell Tank Wagon	£1
R13 Coal Truck	£1
R14 Fish Van	£1
R15 Milk Tank Wagon	£1
R16 Brake Van	£1
R17 Bolster Wagon	£1
R11/A Ventilated Van	£1
R17/C Flat Wagon and Minic Car	£2
R18 Cable Drum Wagon	£1
R19 Flat Wagon	£1
R110 Bogey Bolster Wagon	£1
R111 Hopper Car	£3
R112 Goods Truck	£1
R113 Goods Truck	£1
R114 Box Car	£1
R115 Caboose	£2
R116 Gondola	£1
R117 Oil tanker	£2
R117/CN Oil Tanker Car	£2
R118 Bogey Well Wagon	£1
R118/A Bogey Wagon	£2
R119 Flat Cart	£3
R112 Cattle Wagon	£2
R123 Horse Box	£2
R124 Brake Van	£2
R126 Stock Car	£3
R127 Operating Crane Truck	£3
R128 Operating Helicopter Car	£2
R129 Refrigerator Car	£3
R136 Box Car	£3
R137 Cement Car	£3
R138 Snow Plough	£4
R139 Pickle Car	£5
R210 Shell Wagon	£2
R211 Shell Oil Wagon	£2
R212 Bogie Bolster Wagon	£2
R213 Bogie Well Wagon	£4
R214 Ore Wagon	£1
R215 Bulk Grain Wagon	£1
R214/A Hopper Wagon	£1
R216 Rocket Lauching Wagon	£2
R217 Open Truck	£1
R218 Closed Van	£2
R219 Bogie Brick Wagon	£2
R234 Flat Car	£1
R235 Pulp Wood Car	£2
R236 Depressed Centre Car	£2
R237 Depressed Car With Low Load	£3
R239 Bomb Transporter	£4
R239K Red Arrow Bomb Transporter	£4
R240 Bogie Brick Wagon	£2
R241 Bogie Well Wagon with Tank	£4
R242 Trestrol Wagon	£2
R243 Mineral Wagon	£1
R244 Mineral Wagon with Coal	£1
R245 Open Wagon with Oil Drum	£1
R246 Open Wagon with Timber Load	£1
R247 Bogie Tank Wagon	£2
R248 Sugar Container Wagon	£5
R249 Exploding Car	£2

R250 Rank Flour Mills Wagon £
R262 Continental Guards Van £
R340 Three Containers Wagon £
R341 Searchlight Wagon £
R342 Car Transporter £
R343 Rocket Launcher £
R345 Side Tipping Flat Car £
R346 Crew Repair Wagon £
R347 Engineering Department Wagon £
R348 Giraffe Car £
R349 Bogie Wagon £
R449 Olde Tyme Caboose £
R344 Track Cleaning Car £
R475 Platform Truck with Crane £
R560 Transcontinental Crane Car £
R561 Triang Container Wagon £
R562 Catapult Plane Car £
R563 Bolster Wagon £
R564 Cement Wagon £
R571 G-10 'Q' Car £
R577 Converter Wagon £
R630 P.O.W. Car £
R631 Tank recovery Wagon £
R632 Armoured Rail Car £
R633 Liner Train £
R636 Guards Van £
R639 Sniper Car £
R647 Dewars Grain Wagon £
R648 Johnny Walker Wagon £
R649 Vat 69 Bulk Wagon £
R650 Haig Bulk Wagon £
R666 Cartic Car Carrier £
R668 Bowaters China Clay Wagon £
R725 Command Car £

MOTORWAYS

M1541 Rolls Royce Silver Cloud £
M1542 Jaguar 3.4 £
M1543 Humber Super Snipe £
M1544 Luxury Coach £
M1545 Double Decker Bus £
M1546 Lorry with Bale Load £
M1547 Bedford Lorry and Load £
M1548 Car Transporter £
M1549 Fire Chief's Humber Saloon £
M1550 Fire Engine £
M1551 Shell Tanker £
M1552 Police Car £
M1553 Caravan £
M1554 Trailer £
M1555 Trailer and Boat £
M1556 Mercedes Benz £
M1557 Austin A/40 £
M1558 Mercedes 300 SL £
M1559 E-Type Jaguar £
M1560 Renault Floride £
M1561 Morris 1000 £
M1562 Mobile Bank £
M1563 Securicor Van £
M1564 Steam Lorry £
M1565 Breakdown Lorry £
M1566 Delivery Van £
M1567 3.4. Jaguar £

1568 3.4. Continental Jaguar	£5
1569 Conqueror Tank	£5
1570 Mechanical Horse	£2
1571 Station Trolley	£5
1572 Royal Mail Van	£5
115573 Aston Martin DB4	£5
1574 Porsche Carrera GT	£5
1575 Daimler Special	£5
1576 Ferrari 500 Superfast	£5
1577 Chevrolet Corvette Stingray	£3
1578 Car Water Ferry	£2
1579 Car Water Loading Quay	£3
1580 Car Water Ferry	£5
1581 Aston Martin DB6	£3
1582 E-Type Jaguar 2 x 2	£3
1587 Ford GT Mk II	£3
1588 Ferrari 330-P4	£3
1589 Alfa Romeo T33	£3
1590 Porsche 907	£5

REWAR TRANSPORT VEHICLES

M Minic Ford £100 Saloon	£5
M Ford Light Van	£3
M Minic Ford Royal Mail Van	£3
M Sports Saloon	£5
M Limousine	£5
M Cabriolet	£5
M Town Coupe	£5
M Open Sports Tourer	£5
M Streamline Saloon Closed	£3
M Delivery Lorry	£5
M Tractor	£3
2M Learner's Car	£2
3M Racing Car	£3
4M Streamline Open Saloon	£3
5M Petrol Tank Lorry	£5
5M Caravan, Non-Electric	£2
7M Vauxhall Town Coupe	£5
8M Vauxhall Cabriolet	£5
M Light Tank	£3
M Triang Transport Van	£3
2M Carter Paterson Van	£2
3M Tipper Lorry	£3
4M Luton Transport Van	£3
5M Delivery Lorry with Cases	£2
4M Tractor and Trailer with Cases	£5
7M Police Patrol Car	£5
8M Tourer with Boat on Trailer	£5
9M Traffic Control Car	£3
M Mechanical Horse and Pantechnicon	£2
M Mechanical Horse and Fuel Tanker	£5
2M Dust Cart	£3
3M Steam Roller	£2
4M Tourer with Passengers	£5
5M Rolls-Type Tourer (Non-Electric)	£3
5M Daimler-Type Tourer (Non-Electric)	£3
7M Bentley-Type Tourer	£3
8M Caravan Set	£5
9M Taxi	£5
M Mechanical Horse and Trailer	£5
M Caravan with Electric Light	£3
2M Rolls-Type Sedanca (Non-Electric)	£2
3M Daimler-Type Sedanca (Non-Electric)	£3

44M Traction Engine
45M Bentley-Type Sunshine Saloon
46M Daimler-Type Sunshine Saloon (Non-Electric)
47M Rolls-Type Sunshine Saloon (Non-Electric)
48M Breakdown Lorry and Crane
49M Searchlight Lorry
50ME Rolls-Type Sedanca with Electric
5IME Daimler-Type Sedanca with Electric
52M Single Decker Bus
53M Single Decker V Bus
54M Traction Engine and Trailer
55ME Bentley-Type Tourer with Electric
56ME Rolls-Type Sunshine Saloon with Electric
57ME Bentley-Type Sunshine Saloon with Electric
58ME Daimler-Type Sunshine Saloon with Electric
59ME Caravan Set with Electric
60M Double Decker Bus
61M Double decker Bus 'G'
62M Fire Engine
62ME Fire Engine with Electric
63M Minic Set No. One
64M Minic Set No. Two
65M Construction Set
66M Six-Wheel Army Lorry
67M Farm Lorry
68M Timber Lorry
69M Canvas Tilt Lorry
69MCF Canvas Tilt Lorry in Camouflage
70MCoal Lorry
71M Mechanical Horse and Milk Tanker
75M Mechanical Horse and Barrels
73M Cable Lorry
74M Log Lorry
75M Ambulance
76M Balloon Barrage/Wagon/Trailer
77M Double Decker Trolley Bus
78M Jeep
79M G.W.R. Van
80M L.M.S. Van
81M L.N.E.R. Van
82M S.R. Van
83M Farm Tractor
84M Jeep

POST—WAR NOVELTIES

Pecking Bird
Jack in the Boat
Barnacle Bill
The Prince and Princess
Loch Ness Monster
Electric Railway
Sky Liner Plane
Sky King Airliner
Ladybird
Magic Puffin
Land Yacht
Mouse
Mighty Jabberwock
Grabbing Spider
Nuffield Tractor
Service Station
Three A Service Station
Sherman Tank

Armoured Car	£5
Racing Car	£5
No.2 Racing Car	£5
Musical Saloon	£5
Electric Pathfinder	£5
No.2 Taxi	£5
No.2 Musical Saloon	£5
London Fire Engine	£3
Clockwork Jeep	£2
No.2 Jeep	£2
Car and Caravan	£5
Traction Engine and Trailer	£5
Traction Engine and Tar Barrel	£5
Farm Tractor	£5
Racing Car	£5
Double Decker Money Box	£3
Cream Roller	£3
Bulldozer	£3
Ambulance	£3
London Taxi	£5
Minic Bulldozer No.2	£5
Reverse Steamroller	£5
Tank	£3
Service Station	£5
Garage	£5
Trolley Bus	£5

POST—WAR COMMERCIALS

5-Ton Tipper	£5
No.2 Tipper	£5
Double Decker Bus	£5
Single Decker Bus	£5
Mechanical Horse and Cruiser	£5
Breakdown Truck	£3
Mechanical Horse and Cable Drum Trailer	£5
Morris Light Van	£3
Horse and Watney Trailer	£5
Clockwork Lorry	£3
British Road services Lorry	£3
Austin A40 Van	£2
Morris 10 cwt. Van	£2
Royal Mail Van	£5
Morris Telephone Van	£5
Timber Lorry	£2
Dump Truck	£2
Shutter Van	£3
BR Services Van	£3
Tractor and Trailer with Cases	£2
Mechanical Horse and Log Trailer	£3
Mechanical Horse and Horsebox	£3
Heavy Transport Van	£2
Petrol Tanker	£2
Clockwork Case Lorry	£3
Mechanical Horse and Super Pantechnicon	£3
Large Mechanical Horse and Fuel Tanker	£3
Ice Cream Van	£3
Bakery Van	£3
Luton Clockwork Van	£3
Refuse Lorry	£3
Mechanical Horse and Brewers Trailer	£5
Mechanical Horse and Milk Tanker	£2
Post Office Telephone Van	£2
Removal Van	£2

Friction Morris Minor £
Austin Taxi £
Large Box Van £
Open Lorry £
Routemaster Double Decker Bus £

POST-WAR SPORTS CARS AND SALOONS

No.2 Sports Car with Horn £
Ford Zephyr Saloon £
No.1 Sports Car £
No.2 Sport Car £
Pathfinder Car £
No.2 Ford Zephyr Saloon £
Vanguard Saloon £
No.2 Vanguard Saloon £
Stop-On Saloon £
No.2 Stop-On Saloon £
Police Car £
Mu. Buick Sedan £
Control Special Car £
Rolls Sunshine Saloon £
Jowett Javelin Car £
Riley Saloon £
Ford Monarch Sedan £
Rolls Sedan Car £
Roomy Sports Special £
Rolls Tourer £
Morris Minor £
Minic 'O' Saloon £
Hillman Minx Saloon £
Morris Oxford £
Standard Vanguard £
Streamline Sports Saloon £
Hurricane Armstrong Siddeley £
Rover 90 £

SPOT-ON VEHICLES

100 Ford Zodiac £
101 Armstrong Siddeley Sapphire £
102 Bentley Saloon £
103 Rolls Royce Silver Wraith £
104 MGA Sports Car £
105 Austin Healey £
106 Baby Austin £
107 Jaguar XKSS £
108 Triumph TR3 £
110 Ford Popular £
111 Morris 8 Tourer £
112 Jensen 541 £
113 Aston Martin DB3 £
114 Jaguar 3.4 £
115 Bristol 406 £
117 Sunbeam Motor Cycle and Sidecar £
118 BMW Isetta £
119 Meadows Frisky Sports £
120 Fiat Multipla £
125 Singer Gazelle £
131 Super Goggomobil £
133 Utility Jeep £
154 Austin A40 £
155 Austin Taxi £
157 Rover 3-Litre £

8 Rover	£5
9 Morris Minor	£5
5 Vauxhall Cresta	£5
6 Renualt Floride	£5
8 Vauxhall Victor	£5
3 Humber Super Snipe Estate	£5
4 Austin A60	£3
5 Humber Saloon	£3
6 Austin Estate Car	£5
0 SL Ford Zodiac	£5
1 Sunbeam Alpine	£3
2 Sunbeam Sports	£5
3 NSU Prinz	£3
0 Morris Minor Mini Van	£3
1 Austin Baby Seven	£5
3 Ford Anglia	£3
5 Daimler SP 250	£3
7 E-Type Jaguar	£5
9 Austin Healey Sprite	£5
0 Austin Healey Saloon	£5
0 Royal Rolls Royce	£5
1 Volvo P1800	£3
2 Morris 1100	£2
3 Supercharged Bentley	£5
4 Armstrong Siddeley	£5
6 Bullnose Morris	£5
7 MG 100	£3
8 Model T Ford	£5
9 Zephyr 6 and Caravan	£5
0 Ford Zephyr 6	£3
1 Riley Pathfinder	£5
4 Morris 1100 with Canoe	£5
6 S-Type Jaguar	£5
8 Mercedes 230 SL	£5
0 Vauxhall Cresta	£5
6 Austin 1800	£5
7 Hillman Minx	£5
8 Hillman Estate	£5
7 Volkswagen	£5
9 Zephyr 6 Police Car	£5
23 Riva Speed Boat	£5
3106 4-Wheeled Trailer	£5
6A Austin Type OC 503	£5
8 Daimler Bus	£25
9 Special Box Truck	£15
9/2 Erf Lorry 68G	£5
9/3 Erf flat float with Sides	£5
0 Mammoth Major Cattle Truck	£5
0/2 AEC Mammoth Major	£5
0/2B Mammoth Major 8 Wagon	£10
0/3 AEC Mammoth Major 8	£10
0/4 AEC Mammoth Major Tanker	£15
1a/OT Ford Thames Trader	£10
6 Caterpillar Tractor D9	£5
7 Caterpillar Tractor Crane	£5
8 Ford Thames Van	£5
9 Morris Pick-Up Truck	£5
2 United Dairies Milk Float	£5
3 Bamforth Excavator	£5
5 LGP Sailing Dinghy	£5
7 Massey Harris Tractor	£5
8/1 Combine Harvester	£5
9 Eccles Caravan	£5
5 London Routemaster Bus	£25

156 Mulliner Luxury Coach £
158/A 3C Bedford Low Loader
158/A/02 Bedford 10 Tonner
161 Long Wheel Base Land Rover
207 Wadham Ambulance
210 Morris Mini Minor Van
264 Tourist Caravan
265 Tonibell Ice Cream Van
269 Zephyr 6 and Classic Caravan
271 Express Dairies Milk Van
273 Commer Security Van
308 Landrover and Trailer

CATALOGUES AND LEAFLETS

Publicity Leaflet 1952. Triang Railways £
Publicity Leaflet 1953. Triang Railways £
Publicity Leaflet 1954. Triang Railways £
Triang Railways Catalogue First Edition 1955 £
Triang Railways Catalogue Second Edition 1956 £
Triang Railways Catalogue Third Edition 1957 £
Triang Railways Catalogue Fourth Edition 1958 £
Triang Railways Catalogue Fifth Edition 1959 £
Triang Railways Catalogue Sixth Edition 1960 £
Triang Railways Catalogue Seventh Edition 1961 £
Triang Railways Catalogue Eighth Edition 1962
Triang Railways Catalogue Ninth Edition 1963
Triang Railways Catalogue Tenth Edition 1964 £
Triang Railways Catalogue Eleventh Edition 1965
Triang Railways & Hornby Dublo Amalgamation Leaflet 1965 £
Triang-Hornby Railways Catalogue Twelfth Edition 1966 £
Triang-Hornby Railways Catalogue Thirteenth Edition 1967
Triang-Hornby Railways Catalogue Fourteenth Edition 1968
Triang-Hornby Railways Catalogue Fifteenth Edition 1969
Triang-Hornby Railways Catalogue Sixteenth Edition 1970 (Rare) £
Triang Minic Catalogue 1965. Presented with Minic Sets
Triang Railways Servicing Booklet. Any Edition
Triang Railways Instruction Manual Any Edition

SPOT-ON CATALOGUES

First Edition 1959 £
Second Edition 1960 £
Third Edition 1961 £
Fourth Edition 1962 £
Fifth Edition 1963 £
Sixth Edition 1964 £
Seventh Edition 1965 £
Eighth Edition 1966 £

MECCANO MAGAZINES WITH MINIC ADVERTS.

1933 Edition £
1934 Edition £
1935 Edition £
1936 Edition £
1937 Edition £
1938 Edition £
1939 Edition £
1945 Edition £
1946 Edition £
1947 Edition £
1948 Edition £
1949 Edition £
1950 Edition £
Triang-Minic Lines Brothers General Lists. 1945–1955   £10 ea
Triang-Minic Lines Brothers Toys Booklet. 1950, 1951, 1952, 1953, 1954, 1955, 1956   £10 ea